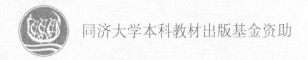
同济大学本科教材出版基金资助

Comprehensive Experiment of Life Science

Chief editor Lyu Lixia

同济大学出版社
TONGJI UNIVERSITY PRESS

Introduction

Life sciences are helpful in improving the quality and standard of life. While biology remains the centerpiece of the life sciences, technological advances in cell biology, molecular biology and biochemistry have led to a burgeoning of specializations and interdisciplinary fields. Practical course is the basis of life science. This English edition of the successful textbook provides readers with a modern and basic experience in experimental biochemistry and cell biology.

The textbook is designed for international medical students and also can be applied in all levels of medical students with English-taught program.

图书在版编目(CIP)数据

生命科学综合实验＝Comprehensive Experiment of Life Science:英文/吕立夏主编. —上海:同济大学出版社, 2017.12
　ISBN 978-7-5608-7121-9

　Ⅰ.①生… Ⅱ.①吕… Ⅲ.①生命科学—实验—教材—英文 Ⅳ.①Q1-0

中国版本图书馆CIP数据核字(2017)第144436号

Comprehensive Experiment of Life Science
Chief editor　Lyu　Lixia

| 责任编辑 | 赵　黎 | 责任校对 | 徐逢乔 | 封面设计 | 陈益平 |

出版发行	同济大学出版社　　www.tongjipress.com.cn
	(地址:上海市四平路1239号　邮编:200092　电话:021-65985622)
经　销	全国各地新华书店
排　版	南京月叶图文制作有限公司
印　刷	当纳利(上海)信息技术有限公司
开　本	787 mm×1092 mm　1/16
印　张	6.5
字　数	162 000
版　次	2017年12月第1版　2017年12月第1次印刷
书　号	ISBN 978-7-5608-7121-9
定　价	60.00元

本书若有印装质量问题,请向本社发行部调换　　版权所有　侵权必究

Comprehensive Experiment of Life Science

Chief editor
Lyu Lixia

Associate editor
Li Jiao   Gao Furong

Editor

Lyu Lixia	Li Jiao	Gao Furong
Jin Caixia	Tian Haibin	Zhang Jieping
Wang Juan	Zhang Jingfa	Guo Feng
Sha Jihong	Shao Zhihua	Xu Jingying
Xu Lei	Shi Xiujuan	Chen Ling
Gao Shane	Cui Yingyu	Zhang Lei
Li Siguang	Xu Guotong	Jia Song
Xu Jie		

Preface

Modern science development heavily depends on interdisciplines. Since 2010, we have integrated experiments of cell biology, biochemistry, molecular biology and genetics and established "Comprehensive experiments of life science" course. The textbook is mainly designed for MBBS students and English-taught Chinese students, and contains validating and comprehensive experiments. Validating experiments include basic cellular and biochemical techniques, namely, observation of cell cycle, cell culture and cell fusion, spectrophotometry, chromatography, centrifugation and electrophoresis. Comprehensive experiments contain three experiments of isolation, and adipogenic differentiation of bone marrow mesenchymal stem cell, polymorphism analysis of angiotensin converting enzyme (ACE) gene and clinical biochemistry of blood glucose and lipid analysis of normal and diabetic rats.

Experiments are bridge linking theory and research and clinical application. After the lab practice, students should be able to master the basic techniques, develop the basic scientific thinking and apply what they have learned from lectures to research and clinical practice.

Every colleague in our team has made his great effort to the experiment design, text writing, and editing. Due to limited time, the writing is not perfect and errors are unavoidable. Suggestions from students and colleagues have been most helpful in the formulation of this edition. We look forwards to receiving similar input in the future.

We thank all our team members' contribution to this textbook. Special thanks are extended to Department of Teaching Affairs Tongji University for providing funding on publishing.

Chief editor
2017/10/9

Preface

Part 1 Biochemistry Experiment 1

Experiment 1　Use of the Spectrophotometer & Determination of Protein Concentration ……………………………………………………… 3

Experiment 2　Chromatography ……………………………………………………… 17

　Experiment 2（1）　Using Gel Filtration Column Chromatography to Separate Hemoglobin and Nucleoprotamine ………………………………… 25

　Experiment 2（2）　Separating Amino Acids Mixture by Lon-Exchange Chromatography …………………………………………………… 30

Experiment 3　Determination of K_m of Alkaline Phosphatase ………………………… 34

Experiment 4　Centrifugation Technique—Adaptive Immune Response: Isolation and Identification of Lymphocyte ………………………………………… 41

　Experiment 4（1）　The Separation of Monocytes by Density Gradient Centrifugation ……………………………………………………………………… 45

　Experiment 4（2）　Lymphocyte Identification——Counting the Number of B Lymphocyte ………………………………………………………… 48

Experiment 5　Polyacrylamide Gel Electrophoresis of Proteins …………………………… 51

Experiment 6　Effect of Hormone on Blood Sugar and Lipoprotein of Normal and Diabetic rat ……………………………………………………………… 59

Experiment 7　Polymorphism Analysis of ACE Gene by Extracting Genome DNA of Buccal Epithelial Cells ………………………………………………… 66

Part 2　Cell Biology Experiment　71

Experiment 8　Optical Microscope and Cell Cycle ················· 73

Experiment 9　Cell Culture and Cell Fusion ························ 84

Experiment 10　Primary Culture and Directional Differentiation of Rat Bone Marrow Mesenchymal Stem Cells ························ 91

Reference　96

Part 1
Biochemistry Experiment

Experiment 1

Use of the Spectrophotometer & Determination of Protein Concentration

1.1 Objectives

(1) Understand the principle of spectrophotometry.
(2) Be able to use the spectrophotometer to determine the protein concentration.
(3) Be familiar with the commonly used methods for protein quantification.

1.2 Principles

1.2.1 Spectrophotometry

Spectrophotometry is a means for determining the concentration of a substance in solution. A dissolved substance will absorb light of specific wavelength (electromagnetic radiation) and thus decrease the amount of light that passes through the sample. By using light of the appropriate wavelength, the concentration of an absorbing substance can be determined by comparing the intensity of the light before and after it passes through the solution.

Electromagnetic radiation occurs in a spectrum of wavelengths that extends from gamma rays of very short wavelengths to radio waves that have wavelengths measured in meters (Figure 1-1). The region of the electromagnetic spectrum that is most useful for the investigation of biological systems lies between wavelengths of about 240 nm and 800 nm. This range includes part of the ultraviolet region (from 190 nm to 380 nm) and the region of visible radiation, called "visible light" (from 380 nm to 750 nm).

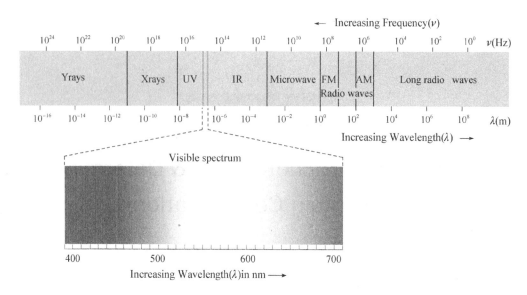

Figure 1-1　Electromagnetic radiation spectrum

When two colors (visible light) are mixed in appropriate proportions, a whitecolor will be produced. Thus, these two colors are called complementary color (Figure 1-2). The color we see in a sample solution is due to the selective absorption of certain wavelengths of visible light and transmittance of the remaining light. If the sample absorbs all wavelengths in the visible region, it will appear black; if it absorbs none of them, it will appear white or colorless.

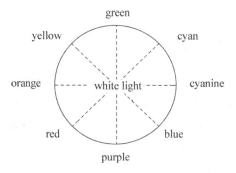

Figure 1-2　Complementary color

If development of color is linked to the concentration of a substance in solution, its concentration can be measured by determining the extent of absorption of light at the specific wavelength. For example, hemoglobin appears red because the hemoglobin absorbs blue and green light rays much more effectively than red. The degree of absorbance of blue or green light is proportional to the concentration of hemoglobin. The visible range is only a very small part of the electromagnetic spectrum. Ultraviolet (UV) and infrared (IR) spectrophotometric methods are suitable for many colorless substances that absorb strongly in the UV and IR spectral regions.

1.2.2 Transmittance, Absorbance and the Lambert-Beer Law

As shown in Figure 1-3, Transmittance (T) is defined as the ratio of the amount of the transmitted light to the amount of the incident light.

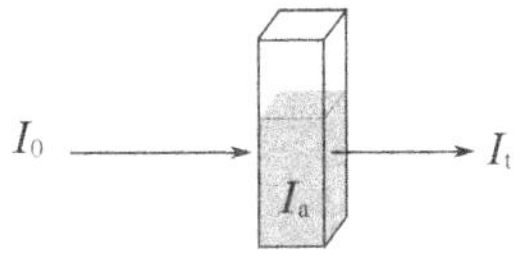

$I_0 = I_a + I_t$
I_0 = intensity of incident light
I_a = intensity of absorbed light
I_t = intensity of transmitted light

Figure 1-3 The relationship between I_0, I_a and I_t

Absorbance (A) is defined as the negative logarithm of the transmittance, and absorbance and transmittance bear an inverse relationship.

$$T = I_t / I_0$$
$$A = -\log T = -\log I_t / I_0$$

In most applications, one wishes to relate the amount of light absorbed to the concentration of the absorbing molecule. It turns out that the absorbance rather than the transmittance is most useful for this purpose.

In 1760, Lambert found the relationship between A and the thickness of liquid layer (L): When light wavelength, solvent, concentration of solution and temperature are fixed, A is only proportional to L.

$$A = -\log I_t / I_0 = k_1 L$$

K_1 is a constant, which is related to substance properties, incident light wavelength, solvent, concentration of solution and temperature.

In 1852, Beer established the equation of A and concentration of solution (C): When wavelength, solvent, thickness of liquid layer and temperature are same, A is only proportional to C.

$$A = -\log I_t / I_0 = k_2 C$$

K_2 is a constant, which is related to substance properties, incident light wavelength, solvent, solution thickness and temperature.

The two equations were combined together and the Lambert-Beer Law was obtained (Figure 1-4):

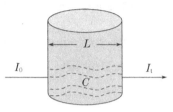

$A = \log I_0/I_t = KCL$
A = amount of light absorbed by the solution expressed as absorbance or optical density
I_0 = intensity of incident light
I_t = intensity of transmitted light
K = constant (extinction coefficient)
C = concentration of the substance
L = path length of the light-absorbing sample

Figure 1-4　The formula of Lambert-Beer Law

1.2.3　Analysis of Sample Concentration

1.2.3.1　Standard Comparison Method

In practice, the concentration of a solute in a sample with unknown concentration or named as unknown sample (u) can be determined directly by comparing the A of the unknown sample to the A of a standard solution whose concentration is known or named as standard sample (s) providing that such compounds obey the Lambert-Beer Law and all conditions under which standards and unknowns prepared should be kept identical.

$$\text{Standard sample } A_s = K_s \times C_s \times L_s$$
$$\text{Unknown sample } A_u = K_u \times C_u \times L_u$$

Since the L, the path length and the extinction coefficient will be constant; that is, $L_s = L_u$ and $K_s = K_u$, thus, $A_s/A_u = C_s/C_u$, $C_u = A_u/A_s \times C_s$.

1.2.3.2　Standard Curve Method

According to the Lambert-Beer Law, there is a linear relationship between absorbance and concentration of a solute when all conditions under which standards and unknowns prepared should be kept identical. So a plot of absorbance vs. concentration of absorbing solute yields a straight line passing through the origin (Figure 1-5). This is usually done by preparing a series of standard solutions, each with a known concentration of a given compound, measuring their absorbance values and plotting absorbance *vs.* concentration to construct a curve. The concentration of the unknown sample can be located by drawing a straight line from point of absorbance of the unknown until it intersects with concentration curve, and then draw perpendicularly to the x-axis to identify the concentration of the unknown sample.

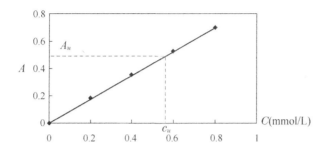

Figure 1-5 Standard curve

The Lambert-Beer Law implies that when concentration is equal to zero ($C = 0$), absorbance must also be zero ($A = 0$). In other words, the standard curve must pass through the origin.

C is the concentration of solute and Aunknown(A_u) is the absorbance of solute under specific wave length.

1.2.4 Parameter Selection

1.2.4.1 Choice of Wavelength

The plot of absorbance of a sample *vs.* wavelengths is called the absorption spectrum. Theoretically we could choose any wavelength for quantitative analysis of concentration. However, the magnitude of the absorbance is important, especially when you are trying to detect very small amounts of substance. For this reason we generally select the wavelength of maximum absorbance for a given sample and use it in our absorbance measurements.

1.2.4.2 Choice of Absorbance

It is strongly recommended to measure absorbance in the range 0.05~1.0 for the following reasons: ① When you are trying to detect very small amounts of substance, the magnitude of the absorbance is important. ② When the absorption band has a "flat" top, the rate of change in absorbance with wavelength is smaller than that on the rising and falling shoulder of the peaks.

1.2.4.3 Blank reference solution

Since transmittance is a relative measurement, the light transmitted by the sample should be compared to the light transmitted by a "reference" solution (Blank). The reference solution is generally the solvent in which the colored compound you are interested in is dissolved. A reference is necessary because the solvent itself might absorb some light at the wavelength you are using and you must correct for that absorbance. We assume that the blank samples transmit 100% of the light entering it that the scale is set to read zero absorbance. Now you can use the full scale of the spectrophotometer.

1.2.5 Spectrophotometer

Aspectrophotometer consists of two instruments, namely a spectrometer for producing light of any selected color (wavelength), and a photometer for measuring the intensity of light(Figure 1-6). The instruments are arranged so that liquid in a cuvette can be placed between the spectrometer beam and the detector. The amount of light passing through the cuvette is measured by the photometer. The photometer delivers a voltage signal to a display device, normally a galvanometer. The signal changes as the amount of light absorbed by the liquid changes.

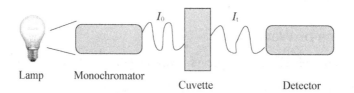

Figure 1-6 Schematic diagram of working principle of a spectrophotometer

1.2.6 Determination of Protein Concentration

1.2.6.1 Bradford Assay

1) Principle

The Bradford assay is a protein determination method that involves the binding of Coomassie Brilliant Blue G-250 dye to proteins. The dye exists in three forms: cationic (red), neutral (green), and anionic (blue). Under acidic conditions, the dye is predominantly in the doubly protonated red cationic form (A_{max} = 470 nm). However, when the dye binds to protein, it is converted to a stable unprotonated blue form (A_{max} = 595 nm). It is this blue protein-dye form that is detected at 595 nm in the assay using a spectrophotometer or microplate reader. Work with synthetic polyamino acids indicates that Coomassie Brilliant Blue G-250 dye binds primarily to basic (especially arginine) and aromatic amino acid residues. Spector (1978) found that the extinction coefficient of a dye-albumin complex solution was constant over a 10-fold concentration range. Thus, Lambert-Beer law may be applied for accurate quantitation of protein by selecting an appropriate ratio of dye volume to sample concentration. Certain chemical-protein and chemical-dye interactions interfere with the assay. Interference from non-protein compounds is due to their ability to shift the equilibrium levels of the dye among the three colored species. Known sources of interference, such as some detergents, flavonoids, and basic protein buffers, stabilize the green neutral dye species by direct binding or by shifting the pH. Nevertheless, many chemical reagents do not directly affect the development of dye color when used in the standard protocol.

2) Procedure

(1) The standard protocol can be performed in three different formats, 5 mL and 1 mL cuvette assay, or a 250 μL microplate assay. The linear range of these assays for albumin (BSA) is 125~1 000 μg/mL, whereas with gamma-globulin the linear range is 125~1 500 μg/mL.

(2) Remove the 1 × dye reagent from 4℃ storage and let it warm to ambient temperature. Invert the 1× dye reagent a few times before use.

(3) If 2 mg/mL BSA standard is used, refer to the below table (Table 1-1) as a guide for diluting the protein standard in test tubes. For the diluent, use the same buffer as in the samples. Protein solutions are normally assayed in duplicate or triplicate. For convenience, blank samples (0 μg/mL) should be made using PBS and dye reagent.

Table 1-1 Preparation of BSA Standards

Reagent	0	1	2	3	4	5	6
2mg/mL BSA (μL)	0	5	10	20	30	60	
PBS(μL)	150	145	140	130	120	90	Sample:150
BSA conc. (mg/mL)	0	0.067	0.13	0.27	0.4	0.8	

(4) Add 2.85 mL of Coomassie Blue to each tube and mix by vortex, or inversion.

(5) Incubate at room temperature for at least 5 min. Samples should not be incubated longer than 1 h at room temperature.

(6) Set the wavelength of spectrophotometer to 595 nm. Zero the instrument with the blank sample. Measure the absorbance of the standards and unknown samples.

(7) Create a standard curve by plotting the 595 nm values (y-axis) versus their concentration in μg/mL (x-axis). Determine the unknown sample concentration by the standard curve. If the samples were diluted, adjust the final concentration of the unknown samples by multiplying the dilution times.

Notes

① Glass or plastic cuvette should be used in the experiment, because Coomassie Brilliant Blue G can firmly bind on quartz cuvette.

② Much more accurate results can be obtained if the standard curve is created by analyzed protein, because proteins have different binding capacity with Coomassie Brilliant Blue G.

③ Standards and samples should be measured according to from low to high concentration order. Measurement should be done continuously. Washing cuvette is not needed for different standards or samples.

3) Results and analysis

Plot standard curve and calculate the protein concentration of unknown sample.

4) Reagents

(1) Bovine serum albumin (BSA): 0.2 g BSA was dissolved in 0.9% NaCl to make the final concentration to be 2 mg/mL.

(2) Coomassie Brilliant Blue G-250: 100mg Coomassie Brilliant Blue G-250 was dissolved in 50 mL 95% ethonal, 120 mL 85% phosphoric acid (H_3PO_4) was added into solution, and then H_2O was added to make the final volume to be 1L.

(3) 0.9% NaCl: 0.9 g NaCl was dissolved in 100 mL H_2O.

(4) PBS: (pH 7.2~7.4): NaCl 137 mmol/L, KCl 2.7 mmol/L, Na_2HPO_4 10 mmol/L, KH_2PO_4 1.76 mmol/L.

5) Instruments

(1) Spectrophotometer.

(2) Micropipettes and tubes.

1.2.6.2 BCA assay

1) Principles

The BCA Protein Assay is a detergent-compatible formulation based on bicinchoninic acid (BCA) for the colorimetric detection and quantitation of total protein. This method combines the well-known reduction of cuprous cation Cu^{2+} to Cu^{+} by protein in an alkaline medium (the biuret reaction) with the highly sensitive and selective colorimetric detection of the Cu^{+} using a unique reagent containing bicinchoninic acid. The purple-colored reaction product of this assay is formed by the chelation of two molecules of BCA with one Cu^{+} (Figure 1-7). This water-soluble complex exhibits a strong absorbance at 562 nm that is nearly linear with increasing protein concentrations over a broad working range (20 ~ 2 000 μg/mL). The BCA method is not a true end-point method, that is, the final color continues to develop. However, following incubation, the rate of continued color development is sufficiently slow to allow large numbers of samples to be assayed together.

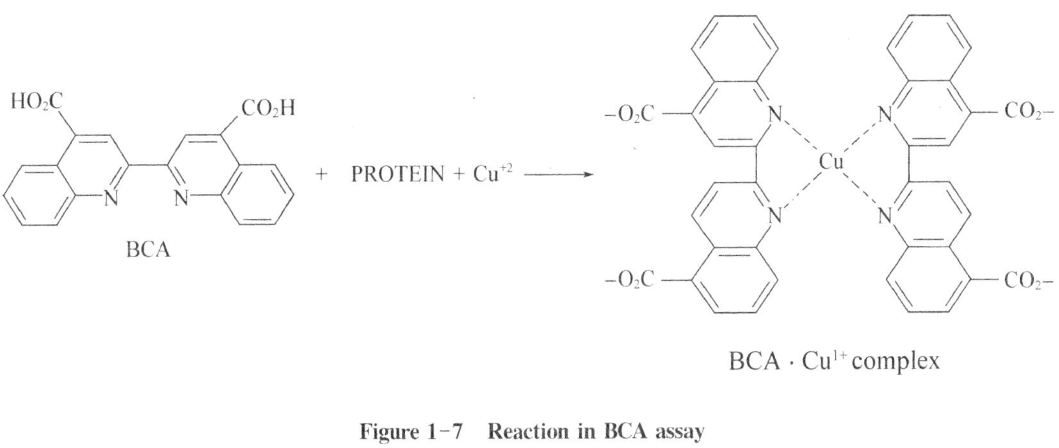

Figure 1-7 Reaction in BCA assay

The macromolecular structure of protein, the number of peptide bonds and the presence of four particular amino acids (cysteine, cystine, tryptophan and tyrosine) are reported to be responsible for color formation with BCA. Studies with di-, tri- and tetrapeptides suggest that the extent of color formation caused by more than the mere sum of individual color-producing functional groups. Accordingly, protein concentrations generally are determined and reported with reference to standards of a common protein such as bovine serum albumin (BSA). A series of dilutions of known concentration are prepared from the protein and assayed alongside the unknown(s) before the concentration of each unknown is determined based on the standard curve. Two assay procedures are presented. The Test Tube Procedure requires a larger volume (0.1 mL) of protein sample, however, because it uses a sample to working reagent ratio of 1 : 20, the effect of interfering substances is minimized. The Microplate Procedure affords the sample handling ease of a microplate and requires a smaller volume (10~25 μL) of protein sample; however, because the sample to working reagent ratio is 1 : 8, it offers less flexibility in overcoming interfering substance concentrations and obtaining low levels of detection.

2) Procedures

(1) Preparation of Diluted Albumin (BSA) Standards

Use Table 1-2 as a guide to prepare a set of protein standards. Dilute the contents of one Albumin Standard (BSA) ampule into several clean vials, preferably using the same diluent as the sample(s). Each 1 mL ampule of 2 mg/mL Albumin Standard is sufficient to prepare a set of diluted standards for either working range suggested in Table 1-2. There will be sufficient volume for three replications of each diluted standard.

Table 1-2 Preparation of Diluted Albumin (BSA) Standards

Vial	ddH$_2$O(μL)	2 mg/mL BSA (μL)	Final BSA Concentration (mg/mL)*
A	0	50 of Standard	2.0
B	125	375 of Standard	1.5
C	325	325 of Standard	1.0
D	175	175 of vial B dilution	0.75
E	325	325 of vial C dilution	0.5
F	325	325 of vial E dilution	0.25
G	325	325 of vial F dilution	0.125
H	400	100 of vial G dilution	0.025
I	400	0	0 = Blank

* Dilution Scheme for Standard Test Tube Protocol and Microplate Procedure (Working Range = 20 ~ 2,000 μg/mL)

(2) Microplate Procedure (Sample : BCA working solution = 1 : 8)

① Pipette 25 μL of each standard or unknown sample replicate into a microplate well (working range = 20~2 000 μg/mL).

Note:

If sample size is limited, 10 μL of each unknown sample and standard can be used (Sample : BCA working solution=1 : 20). However, the working range of the assay in this case will be limited to 125~2 000 μg/mL.

② Add 200 μL of the BCA working solution to each well and mix plate thoroughly on a plate shaker for 30 seconds.

③ Cover plate and incubate at 37°C for 30 minutes.

④ Cool plate to RT. Measure the absorbance at or near 562 nm on a plate reader.

Notes:

① Wavelengths from 540~590 nm have been used successfully with this method.

② Because microplate readers use a shorter light path length than cuvette spectrophotometers, the Microplate Procedure requires a greater sample to working reagent(WR) ratio to obtain the same sensitivity as the standard Test Tube Procedure. If higher 562 nm measurements are desired, increase the incubation time to 2 hours.

③ Increasing the incubation time or ratio of sample volume to WR increases the net 562 nm measurement for each well and lowers both the minimum detection level of the reagent and the working range of the assay. As long as all standards and unknowns are treated identically, such modifications may be useful.

④ Subtract the average 562 nm absorbance measurement of the Blank standard replicates from the 562nm measurements of all other individual standard and unknown sample replicates.

⑤ Prepare a standard curve by plotting the average Blank-corrected 562 nm measurement for each BSA standard vs. its concentration in μg/mL. Use the standard curve to determine the protein concentration of each unknown sample.

⑥ If using curve-fitting algorithms associated with a microplate reader, a four-parameter (quadratic) or best-fit curve will provide more accurate results than a purely linear fit. If plotting results by hand, a point-to-point curve is preferable to a linear fit to the standard points.

3) Results and analysis

Plot standard curve and calculate the protein concentration of the unknown sample.

4) Reagents

(1) BCA reagent A: 1% BCA, 2% $NaCO_3$, 0.16% sodium tartrate, 0.4% NaOH, 0.95% $NaHCO_3$, adjust pH to be 11.25.

(2) BCA reagent B: 4% $CuSO_4$.

(3) 0.9% NaCl: 0.9 g NaCl was dissolved in 100 mL H_2O.

(4) PBS: (pH7.2~7.4): NaCl 137 mmol/L, KCl 2.7 mmol/L, Na_2HPO_4 10 mmol/L, KH_2PO_4 1.76 mmol/L.

5) Instruments

(1) Spectrophotometer.

(2) Micropipettes and tubes.

(3) Microplate.

(4) Thermostat water bath.

1.2.6.3 Biuret Assay

1) Principle

The Biuret reaction can be used for both qualitative and quantitative analysis of protein. The biuret method depends on the presence of peptide bonds in proteins. When a solution of proteins is treated with cupric ions (Cu^{2+}) in a moderately alkaline medium, a purple colored Cu^{2+}-peptide complex (Figure 1-8) is formed which can be measured quantitatively by spectrophotometer in the visible region. So, Biuret reagent is alkaline copper sulfate solution.

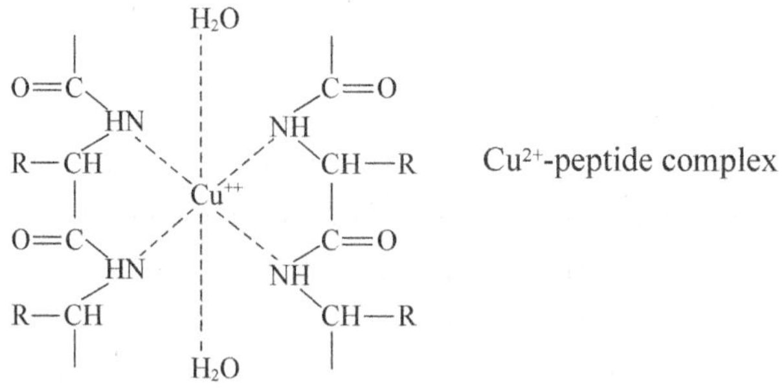

Figure 1-8 Reaction in biuret assay

The intensity of the color produced is proportional to the number of peptide bonds that are reacting, and therefore to the number of protein molecules present in the reaction system. The reaction dosn't occur to amino acids because the absence of peptide bonds, and also that with di-peptide because presence of only one peptide bond, but do with tri-, oligo-, and poly-peptides. Biuret reaction needs presence of at least two peptide bonds in a molecule. The reaction occurs with any compound containing at least two bonds of: —HN—CO—, —HN—CH_2— and

—HN—CS—. The reaction takes its name "Biuret Reaction" from the fact that biuret itself, obtained by heating urea, gives a similar colored complex with cupric ions.

2) Procedures

(1) Label 9 test tubes as (1 to 9) and place them in a test tube rack.

(2) Add to each tube the solutions in the following Table 1-3.

Table 1-3 Preparation of Biuret Standards.

Reagents (mL)	0	1	2	3	4	5	6
2 mg/mL BSA	—	0.3	0.6	1.2	1.8	2.4	3.0
Distilled water	3	2.7	2.4	1.8	1.2	0.6	—
BSA concentration (mg/mL)	0	0.2	0.4	0.8	1.2	1.6	2.0

(3) Add 3.0 mL Biuret reagent to each tube.

(4) Mix well by vortex mixer and incubate at 37℃ for 30 minutes.

(5) Read the absorbance for each tube against the blank at 540 nm.

(6) Calculate the protein concentration in each tube of standard.

(7) Plot the standard curve using concentration of standard tubes of BSA (μg/mL).

(8) Calculate the mean of absorbance of the duplicate sample and obtain the concentration of protein in the sample from the standard curve.

3) Results and analysis

Plot standard curve and calculate the protein concentration of the unknown sample.

4) Reagents

(1) Biuret reagent: 1.5 g $CuSO_4 \cdot 5H_2O$, 6.0 g potassium sodium tartrate, 300 mL 10% NaOH, 1.0 g KI, add H_2O to 1 L. Put solution in brown bottle and if dark red deposition appears, the reagent can not be used.

(2) 0.9% NaCl: 0.9 g NaCl was dissolved in 100 mL H_2O.

(3) PBS: (pH7.2~7.4): NaCl 137 mmol/L, KCl 2.7 mmol/L, Na_2HPO_4 10 mmol/L, KH_2PO_4 1.76 mmol/L.

5) Instruments

(1) Spectrophotometer.

(2) Micropipettes and tubes.

(3) Thermostat water bath.

6) Thought Questions

(1) Explain why you measure the absorbance at a certain wavelength in these experiments? And what type of cuvette did you use?

(2) Describe the alternative method to determine the protein concentration in a given sample in case of not applying standard curve in biuret assay?

(3) Compare the advantages and disadvantages for the three protein quantification methods: BCA, Bradford and Biuret assay.

Experiment 2

Chromatography

Chromatography is the collective term for a set of laboratory techniques used to separate and or to analyze complex mixtures. The mixture is dissolved in a fluid called the "mobile phase", which carries it through a structure holding another material called the "stationary phase".

The various constituents of the mixture travel at different speeds, causing them to separate. The separation is based on differential partitioning between the mobile and stationary phases. Subtle differences in a compound's partition coefficient result in differential retention on the stationary phase and thuschanging the separation. As a result of these differences in mobility, sample components will become separated from each other as they travel through the stationary phase.

2.1 Chromatography Classification

2.1.1 Based on Principle

① Adsorption chromatography: utilizes a mobile liquid or gaseous phase that is adsorbed onto the surface of a stationary solid phase. The equilibriation between the mobile and stationary phase accounts for the separation of different solutes.

② Partition chromatography: It is a liquid-liquid extraction which involves two solvents: liquidstationary phase & a liquid mobile phase.

③ Ion chromatography: Ion chromatography (or ion-exchange chromatography) is a process that allows the separation of ions and polar molecules based on their charge.

④ Gel chromatography: Gel permeation chromatography (GPC) is a type of size exclusion chromatography (SEC), that separates analytes on the basis of size.

⑤ Affinity chromatography: Affinity chromatography is a method of separating biochemical mixtures and based on a highly specific interaction such as that between antigen and antibody.

2.1.2 Based on mode of operation

① Paper chromatography: Paper chromatography is an analytical method technique for separating and identifying mixtures that are or can be colored, especially pigments.

② Thin layer chromatography: Thin-Layer Chromatography (TLC) is a simple and inexpensive technique that is often used to judge the purity of a synthesized compound or to indicate the extent of progress of a chemical reaction.

③ Column chromatography: A separation technique in which the stationary bed is within a tube. The particles of the solid stationary phase or support coated with a liquid stationary phase may fill the whole inside volume of the tube (packed column) or be concentrated on or along the inside tube wall leaving an open, unrestricted path for the mobile phase in the middle part of the tube (open-tubular column).

2.2 Terms in Chromatography

2.2.1 Retention Value

Theretention factor (R_f) Values Chromatography is defined as a relative value that indicates the mobility of the compound with respect to solid phases. R_f values can be calculated as the mobility distance of the sample from its stationary position divided by the distance covered by the solvent solution.

2.2.2 Peak Width

There are three kinds of peak width (W), as listed in Figure 2-1:

The peak width (W) is the distance between each side of a peak measure at 0.607 of the peak height ($0.607h$). The peak width measured at this height is equivalent to two standard deviations (2σ) of the Gaussian curve and thus has significance when dealing with chromatography theory.

The peak width at half height ($W/2$) is the distance between each side of a peak measured at half the peak height ($0.5h$), $W/2 = 2.345\sigma$. The peak width measured at half height has no significance with respect to chromatography theory.

The peak width at the base (W_b) is the distance between the intersections of the tangents drawn to the sides of the peak and the peak base geometrically produced. The peak width at the base is equivalent to four standard deviations (4σ) of the Gaussian curve and thus also has significance when dealing with chromatography theory. $W_b = 4\sigma$.

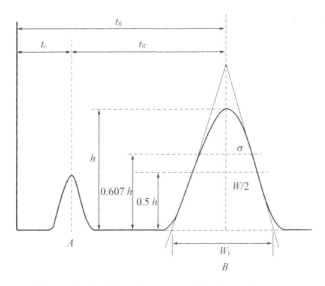

Figure 2-1　Schematic diagram for chromatography

2.2.3　Peak Resolution (R_s)

Resolution (R_s) shows the ability to distinguish peaks in chromatography.

$$R_s = \frac{t_{R_2} - t_{R_1}}{\frac{1}{2}(W_1 + W_2)}$$

During separation, different content of different components results indifferent peak area and peak width, so test for perpendicular line is commonly used to calculate R_s. As shown in Figure 2-2, joint summits of two adjacent peaks and make a vertical line from valley between peaks, distance between the intersection point to peak valley is f and to the base line is g.

$$R_s = \frac{f}{g}$$

If $R_s = 1$, the different components are separated completely and if $R_s < 1$, there is a partial separation; if $R_s \ll 1$, there is almost no separation of different components.

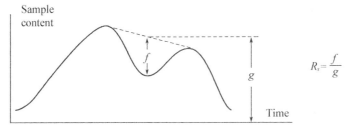

Figure 2-2　Resolution of chromatography

2.2.4 Partition Coefficient (K)

The ratio of concentrations of a compound in the two phases of a mixture of two immiscible solvents at equilibrium.

$$K = \frac{\text{the sample concentrations in the stationary phase}}{\text{the sample concentrations in the mobile phase}}$$

2.2.5 Column Efficiency

The efficiency of a chromatographic column is a measure of the capacity of the column to restrain peak dispersion and thus, provide high resolution. The higher the efficiency, the more the peak dispersion is restrained, and the better the column. The column efficiency will vary with the retention of the peak. In capillary columns, the efficiency generally falls as the retention increases and for a packed column the efficiency generally increases with retention. The expression for calculating the column efficiency can be derived from the plate theory. Column efficiency is measured in theoretical plates (from the Plate Theory) and is taken as 16 times the square of the ratio of the retention distance (the distance between the injection point and the peak maximum) to the peak width at the points of inflection (the points of inflection occur at 0.607 of the peak height). The height equivalent to the theoretical plate (HETP) or the variance per unit length of a column is calculated as the ratio of the column length to the column efficiency.

2.3 Paper Chromatography

Paper chromatography is an analytical technique for separating and identifying mixtures that are or can be colored, especially pigments. This also can be used in secondary or primary colors in ink experiments. This method has been largely replaced by thin layer chromatography, however it is still a powerful teaching tool. Two-way paper chromatography, also called two-dimensional chromatography, involves using two solvents and rotating the paper 90° in between. This is useful for separating complex mixtures of similar compounds, for example, amino acids.

Separations in paper chromatography involve the same principles as those in thin layer chromatography. In paper chromatography, like thin layer chromatography, substances are distributed between a stationary phase and a mobile phase. The stationary phase is usually a piece of high quality filter paper. The mobile phase is a developing solution that travels up the stationary phase, carrying the samples with it. Components of the sample will separate readily according to how strongly they adsorb on the stationary phase versus how readily they dissolve in the mobile phase.

R_f value: The retention factor (R_f) may be defined as the ratio of the distance traveled by the substance to the distance traveled by the solvent. R_f values are usually expressed as a fraction of two decimal places but it was suggested by Smith that a percentage figure should be used instead. If R_f value of a solution is zero, the solute remains in the stationary phase and thus it is immobile. If R_f value = 1 then the solute has no affinity for the stationary phase and travels with the solvent front. To calculate the R_f value, take the distance traveled by the substance divided by the distance traveled by the solvent (as mentioned earlier in terms of ratios). For example, if a compound travels 2.1 cm and the solvent front travels 2.8 cm, (2.1/2.8) the R_f value = 0.75.

2.4 Thin Layer Chromatography

Thin-Layer Chromatography (TLC) is a simple and inexpensive technique that is often used to judge the purity of a synthesized compound or to indicate the extent of progress of a chemical reaction. In this technique, a small quantity of a solution of the mixture to be analyzed is deposited as a small spot on a TLC plate, which consists of a thin layer of silica gel (SiO_2) or alumina (Al_2O_3) coated on a glass or plastic sheet. The plate constitutes the stationary phase. The sheet is then placed in a chamber containing a small amount of solvent, which is the mobile phase. The solvent gradually moves up the plate via capillary action, and it carries the deposited substances along with it at different rates. The desired result is that each component of the deposited mixture is moved a different distance up the plate by the solvent. The components then appear as a series of spots at different locations up the plate. Substances can be identified from their so-called R_f values. The R_f value for a substance is the ratio of the distance that the substance travels to the distance that the solvent travels up the plate. For example, an R_f value of 0.5 means that the spot corresponding to the substance travels exactly half as far as the solvent travels along the plate.

2.4.1 Stationary Phase

As stationary phase, a special finely ground matrix (silica gel, alumina, or similar material) is coated on a glass plate, a metal or a plastic film as a thin layer (~0.25 mm). In addition a binder like gypsum is mixed into the stationary phase to make it stick better to the slide. In many cases, a fluorescent powder is mixed into the stationary phase to simplify the visualization later on (e.g. bright green when you expose it to 254 nm UV light).

2.4.2 Analysis

The components, visible as separated spots, are identified by comparing the distances they have traveled with those of the known reference materials. Measure the distance of the start line to the solvent front (= d). Then measure the distance of center of the spot to the start line (= a). Divide the distance the solvent moved by the distance the individual spot moved. The resulting ratio is called R_f-value. The value should be between 0.0 (spot did not moved from starting line) and 1.0 (spot moved with solvent front) and is unitless.

The R_f (= retardation factor) depends on the following parameters:

① solvent system.

② absorbent (grain size, water content, thickness).

③ amount of material spotted.

④ temperature.

2.5 Ion Chromatography

Ion exchange chromatography (usually referred to as ion chromatography) uses an ion exchange mechanism to separate analytes based on their respective charges. It is usually performed in columns but can also be useful in planar mode. Ion exchange chromatography uses a charged stationary phase to separate charged compounds including anions, cations, amino acids, peptides, and proteins. In conventional methods, the stationary phase is an ion exchange resin that carries charged functional groups which interact with oppositely charged groups of the compound to be retained. Ion exchange chromatography is commonly used to purify proteins using fast protein liquid chromatography.

The ion chromatography is used for analysis of aqueous samples in parts-per-million (ppm) quantities of common anions (such as fluoride, chloride, nitrite, nitrate, and sulfate and common cations like lithium, sodium, ammonium, and potassium) using conductivity detectors. The chromatography also has the capability to analyze aqueous samples for parts-per-billion (ppb) quantities of hydrazine, monomethylhydrazine (MMH), and unsymmetrical dimethylhydrazine (UDMH).

Ion chromatography is a form of liquid chromatography that uses ion-exchange resins to separate atomic or molecular ions based on their interaction with the resin. Its greatest utility is for analysis of anions for which there are no other rapid analytical methods. It is also commonly used for cations and biochemical species such as amino acids and proteins. Most ion-exchange separations are done with pumps and metal columns.

2.6 Gel Chromatography

Size-exclusion chromatography (SEC) is also known as gel permeation chromatography (GPC) or gel filtration chromatography and separates molecules according to their size. Smaller molecules are able to enter the pores of the media and, therefore, molecules are trapped and removed from the flow of the mobile phase. The average residence time in the pores depends upon the effective size of the analyte molecules. However, molecules that are larger than the average pore size of the packing are excluded and thus suffer essentially no retention; such species are the first to be eluted. It is generally a low-resolution chromatography technique and thus it is often reserved for the final "polishing" step of purification.

An extensive used gel is Sephadex. Sephadex is a trademark for cross-linked dextran gel used for gel filtration. By varying the degree of cross-linking, the fractionation properties of the gel can be altered. These highly specialized gel filtration and chromatographic media are composed of macroscopic beads synthetically derived from the polysaccharide, dextran. The organic chains are cross-linked to give a three dimensional network having functional ionic groups attached by ether linkages to glucose units of the polysaccharide chain.

2.7 Affinity Chromatography

Affinity chromatography is a method of separating biochemical mixtures and based on a highly specific interaction such as that between antigen and antibody, enzyme and substrate, or receptor and ligand.

Affinity chromatography can be used to:

① Purify and concentrate a substance from a mixture into a buffering solution.

② Reduce the amount of a substance in a mixture.

③ Discern what biological compounds bind to a particular substance.

④ Purify and concentrate an enzyme solution.

The immobile phase is typically a gel matrix, often of agarose; a linear sugar molecule derived from algae. Usually the starting point is an undefined heterogeneous group of molecules in solution, such as a cell lysate, growth medium or blood serum. The molecule of interest will have a well known and defined property which can be exploited during the affinity purification process. The process itself can be thought of as an entrapment, with the target molecule becoming trapped on a solid or stationary phase or medium. The other molecules in solution will not become trapped as they do not possess this property. The solid medium can then be removed from the mixture, washed and the target molecule released from the entrapment in a process known as elution. Possibly the most common use of affinity chromatography is for the purification of recombinant proteins.

Experiment 2(1) Using Gel Filtration Column Chromatography to Separate Hemoglobin and Nucleoprotamine

2(1).1 Objectives

(1) Understand the principles of gel filtration column chromatography.

(2) Be able to separate hemoglobin and protamines by gel filtration chromatography.

2(1).2 Principle

Protein mixture of different molecular weight is separated by the Sephadex G-50. Protein mixture contains hemoglobin and nucleoprotamine. The molecular weights are 64500 and 2 000~12 000 respectively. The nucleoprotamine will show yellow after pre-colour by dinitrofluorobenzene. Because of the different color and molecular weight, it can be seen that hemoglobin and nucleoprotamine are separated gradually in the chromatographic column and eluted successively.

2(1).3 Procedures

2(1).3.1 Filling the Column

Take a clean glass chromatographic column (0.8~1.5 cm in diameter, 17~20 cm in length) and make clear that the positions of entry and exit. The middle of filling cap of the entry is a little bitinward protuberance and there is a filter screen in the filling cap of the exit. After tightening the filling cap of the exit, please install the column in the vertical position, then add a little water from the entry and check if the water effluxes smoothly from the exit as well as if any leakage around the filling cap. After that, lift the plastic tube of the exit till the same altitude of the entry in order to stop the effusion of the water phase. Open the filling cap of the entry, add the gel suspension of Sephadex from the entry and stir it gently at the same moment. Open the filling cap of the exit and let the water efflux when the gel in the bottom deposits. Keep adding the gel till the height of gel is about 3 cm away from the entry of the column.

The speed of adding gel should be even so that gel may sink evenly, which could prevent the column bed from segmentation and bubble in the column. If the surface of column bed was not smooth, you could stir the gel surface with the fine glass rod to make the gel deposits naturally and the surface of the gel will be smooth eventually.

Notes:
During the process of filling column and separating the protein mixture later, the water level should be always above or as same as the gel level to prevent the dry and cracked gel.

2(1).3.2 Loading Samples

Firstly, open the filling cap of the exit (put down the plastic tube of exit), let the distilled water efflux till the water level as the same as the gel level. Add the protein mixture onto the surface of the gel slowly alongside of inner wall of column with the pipette. Try not to stir the gel bed. Then open the exit, make the samples enter the column bed till the sample phase is at the same level of the gel phase. Add distilled water into the column with same method above till the water level is 2~3 cm higher than the gel bed level. Insert the plastic pipe of entry into the bottle of distilled water; eliminate the bubble in the pipe. Then tighten the cap and begin to elute.

2(1).3.3 Collecting the Samples

Adjust the fluid speed from the exit about 10~15 drops/min, collect effluent with a small beaker, then begin to elute the samples. During the process of elution, you can see that the yellow and red stripes separate gradually, and are eluted successively. Collect eluent of two protein samples respectively.

2(1).4　Results and Analysis

2(1).4.1　Calculate the Recovery Rate

Take 0.1 mL of hemoglobin samples, and add 5 mL distilled water into the hemoglobin solution as standard sample.

Take 5 mL of eluent hemoglobin samples (add distilled water till 5 mL if there's not enough eluent hemoglobin samples.) as measuring sample. Zeroing with distilled water, read the value of the optical density, and calculate the value of recovery rate according to the following formula:

Recovery rate of the hemoglobin (%) = the optical density value of eluent/the optical density value of standard sample × 100%

2(1).4.2　Draw the Chromatographic Profile

The number of collection tube as abscissa and the optical density value of samples as ordinate, then draw a chromatographic profile to observe resolution (R_s).

1) Reagents

(1) Preparation of the gel

Take 1 g of Sephadex G-50, and put it into conical flask. Then add 30 mL of distilled water, place it in the room temperature for 3 hours (room temperature swelling method), or heat it in the boiling water for 1 hour (heat-up swelling method) and place it to room temperature before filling the column.

(2) Preparation of samples

① Preparation of Hb solution

Take 2 mL of Anticoagulation blood into centrifuge tube, 2 500 rpm/min, 5 min, get rid of upper plasma. Wash blood cells with 0.9% Nacl, repeat 3 times. Stir the blood cells gently every time, 3 500 rpm/min, 5 min, get rid of supernate. Add 5 mL of distilled water, mix the sample sufficiently, put it in the fridge overnight in order to hemolyze the blood cells fully.

② Preparation of DNP-nucleoprotamine

Take 0.15 g of nucleoprotamine, dissolve it in the 1.5 mL NaHCO$_3$ solution (pH 8.5~9.0).

Take 0.15 g of dinitrofluorobenzene, dissolve it in the 3 mL of warm 95% ethanol. Add this solution into the nucleprotamine solution immediately when dinitrofluorobenzene dissolves in the ethanol fully.

Mix the above solutions together, then bath it in the boiling water for 5 min. take it out and wait for cooling down. After that, add twice volume (compared to the two solution mixture) of 95% ethanol into the nucleprotamine and dinitrofluorobenzene mixture. The yellow deposit of DNP- nucleprotamine can be observed. Centrifuge it for 5 min and get ride of supernate. Wash the deposit twice with the 95% ethanol.

Dissolve the deposit in 1 mL of distilled water.

③ Preparation of protein mixture solution

Mix 0.1 mL of hemoglobin solution with 0.2 mL of nucleprotamine solution fully.

(3) Other reagents

① Sephdex G-50.

② Anticoagulant blood.

③ 0.9% NaCl.

④ Nucleprotamine.

⑤ 10% NaHCO$_3$.

⑥ 2, 4-dinitrofluorobenzene.

⑦ 95% ethanol.

2) Instruments

(1) Centrifuge.

(2) Filter paper.

(3) Chromatographic column.

(4) Dropper, pipette, test tube.

(5) Beaker.

(6) Spectrophotometer.

3) Thought Questions

(1) Why gel filtration chromatography can separate hemoglobin and nucleoprotamine?

Which one will be eluted first?

(2) How to select gel when perform gel filtration chromatography to separate different components?

Experiment 2 (2) Separating Amino Acids Mixture by Lon-Exchange Chromatography

2(2).1 Objectives

(1) Understand the principles of Ion-exchange Chromatography.
(2) Be able to use the method of separating amino acids mixture.

2(2).2 Principles

In this method, mixture of acidic amino acid (aspartic acid) and alkaline amino acid (lysine) is separated by sulfonic acid ion exchange resin (Dowex 50), the resin contains sulfonic acid group which can be combined with cations. In certain specific conditions, neither their ways of dissociation nor electrical properties are the same. Lysine is positively charged and can be combined with sulfonic resin, while aspartic acid is negatively charged and can not be combined with resin and it can be eluted by effluent. Negatively charged, lysine can not be combined with resin in alkaline conditions and it is eluted. Therefore, during the experiment, the two amino acids can be eluted and separated by means of modification of pH value.

2(2).3 Procedures

2(2).3.1 Filling Column

Take a clean glass chromatography column, and make clear that the position of entry and exit. The middle of filling cap of the entry is a little bit inward protuberance and there is a filter screen in the filling cap of the exit. After tightening the filling cap of the exit, please install the column in the vertical position, then add a little citric acid buffer (pH 4.2) from the entry and check if the water effluxes smoothly from the exit as well as if any leakage around the filling cap. After that, lift the plastic tube of the exit till the same altitude of the entry in order to stop the effusion of the water phase. Open the filling cap of the entry, add the gel suspension of Sephadex form the entry and stir it gently at the same moment. Open the filling cap of the exit and let the water efflux when the gel in the bottom deposits. Keep adding the gel till the height of gel is about 3 cm awat from the entry of the column. The speed of adding gel should be even so that gel may sink evenly, which could prevent the column bed from segmentation and bubble in the column. If the surface of column bed was not smooth, you could stir the gel surface with the fine glass rod to make the gel deposits naturally and the surface of the gel will be smooth eventually.

Notes:
During the process of filling column and separating the amino acids mixture later, the water level should be always above or as same as the gel level to prevent the dry and cracked gel.

2(2).3.2 Loading Samples, Eluting and Collection of Effluent

Firstly, open the cap of the exit, let the buffer solution run off till the solution level and the gel level share the same height, then block the exit. Add themixture of amino acids (0.2 mL) alongside the inner side of chromatography column onto the surface of the gel. Try not to stir the gel bed. Then open the exit, make the samples enter the column bed till the sample phase is at the same level of the gel phase. Add the buffer solution (citric acid, pH 4.2, 3 mL) with the same method above to wash the inner side of the column. When the water level is 1~2 cm higher than the gel level, insert the plastic pipe of entry into the bottle of the buffer solution (citric, pH 4.2), eliminate the bubble in the pipe, tighten the cap, then start the continuous elution. Adjust the fluid speed from the exit about 10~15 drops/min. Collect the effluent with test tubes at the exit (2 mL/tube), collect 6~7 tubes of effluent. Block the exit when finished (for the moment) and verify if the amino acids are eluted. (The method will be introduced in the "Results analysis"part)

If the amino acids are eluted, then remove the reagent bottle of buffer solution (citric acid, pH 4.2), open the exit till both levels share the same height. Add NaOH (0.1 mol/L,

3 mL) and connect the chromatography column to a reagent bottle which contains NaOH (0.1 mol/L) with the hose of the entry. Elute continuously, [same operation as the elution of buffer solution (citric acid)], collect 6~8 tubes, block the exit for the moment and verify if the amino acids are eluted.

After eluting all the amino acids, wash the resin with buffer solution (citric acid, pH 4.2) in order to regenerate the resin which can be reused after regeneration.

(1) Results and Analysis

① Coloration: Number the collecting tubes in accordance with the collecting order, add buffer solution (acetocopal, 0.2 mol/L, pH 5, 0.5 mL) in each tube and blend. Then add ninhydrin solution (0.5%, 0.5 mL). Boil the tubes in a boiling water bath for 10 minutes; the solution's turning into purple blue shows the elution of amino acid. Measure the Absorbance value at the 570nm spot.

② Plot the chromatogram: The number of collection tube as abscissa and the optical density value of samples as ordinate, then draw a chromatographicprofile to observe resolution (Rs).

(2) Reagents

① The preparation of resin: Put the resin (10 g) in the beaker (100 mL), add HCl (2 mol/L, 25 mL), stir for 2 hours, discard the acid solution, and wash the resin with distilled water till neuter. Then add NaOH (2 mol/L, 25 mL), stir for 2 hours, discard the alkaline solution, wash the resin with distilled water till neuter. Conserve the resin in the buffer solution (citric acid pH 4.2).

② Sulfonic acid ion exchange resin (Dower 50).

③ 2 mol/L HCl.

④ 2 mol/L NaOH.

⑤ 0.1 mol/L buffer solution (citric acid, pH 4.2): 0.1 mol/L citric acid 54 mL, 0.1 mol/L sodium citrate 46 mL.

⑥ 0.1 mol/L HCl.

⑦ 0.1 mol/ NaOH.

⑧ 0.2 mol/L buffer solution (acetocopal, pH 5.0): 0.2 mol/L acetocopal 70 mL, 0.2 mol/L sodium acetate 30 mL.

⑨ 0.5% ninhydrin solution: ninhydrin 0.5 g, water 100 mL.

⑩ Mixture of amino acid: Asp (10 mg), Lye (10 mg) dissolving in 0.1 mol/L HCl (1 mL).

(3) Instruments

① Chromatography column.

② Measuring pipette.

③ Reagent bottle.

④ Measuring cylinder.

⑤ Boiling water bath.

⑥ Glass test tube.

⑦ pH test paper.

(4) Thought Questions

Design an experiment to separate a mixture of components with different pI.

Experiment 3

Determination of K_m of Alkaline Phosphatase

3.1 Objectives

(1) Understand the significance of K_m and learn how to determine the K_m value of AKP.

(2) Be able tocalculate the K_m of enzyme using the standard curve.

3.2 Principles

One of the most useful models in the systematic investigation of enzyme rates was proposed by Leonor Michaelis and Maud Menten in 1913. The concept of the enzyme-substrate complex is the central to Michaelis-Menten kinetics. When the substrate(S) binds in the active site of enzyme (E), intermediate complex (ES) is formed. This complex must pass to the transition state (ES $*$); and the transition state complex must advance to an enzyme product complex (EP). The latter is finally competent to dissociate to product and free enzyme. The series of events can be shown as:

$$E + S \longleftrightarrow ES \longleftrightarrow EP \longleftrightarrow E + P$$

Between the binding of substrate to enzyme, and the reappearance of free enzyme and product, a series of complex events must take place. At a minimum an ES complex must be formed.

The relationship between $[S]$ and Velocity (V) can be described by Michaelis-Menten equation, where K_m, referred to as Michaelis constant is introduced. The experimentally determined value K_m is considered a constant that is characteristic of the enzyme and the substrate under specified conditions.

$$V_1 = \frac{V_{max}[S]}{K_m + [S]}$$

V_{max} = maximum velocity that the reaction can attain.

The K_m value is a characteristic constant of enzymes. The K_m value for an enzyme depends on particular substrate and on the environmental conditions, such as temperature, pH, and ionic strength, regardless of enzyme concentration. The lower value of K_m, the greater the affinity of the enzyme for ES complex formation.

Under specified environmental conditions (fixed pH, temperature, ionic strength and enzyme concentration), the relationship between reaction rate [V] and substrate concentration [S] changes as follows (Figure 3-1):

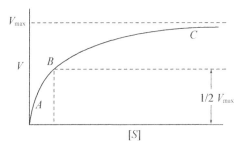

Figure 3-1 Relationship between reaction rate(V) and substrate concentration[S] in a typical enzyme reaction

K_m is the substrate concentration at which the reaction rate is half-maximal. Determination of K_m is an important way to study the enzyme's kinetics. Most of the enzymes have a K_m between 0.01 mol/L and 100 mol/L.

The properties of hyperbolically shaped curve in Figure 3-1:

① At low concentrations of the substrate [S], velocity (V) is proportional to [S], which is a usual features of a first order reaction.

② With the increase of substrate concentration, V does not increase proportionally to [S]. At substrate concentrations far higher than K_m, ([S]≫K_m), the curve approaches to a plateau. The reaction becomes nearly independent of the substrate concentration and shows zero-order kinetics.

③ When the enzyme is saturated by substrate, almost enzyme molecules are present as enzyme substrate complex and the reaction is limited no longer by substrate availability but by the amount and the turnover number of the enzyme.

To avoid dealing with curvilinear plots of enzyme catalyzed reactions, biochemists Lineweaver and Burk introduced an analysis of enzyme kinetics based on the following rearrangement of the Michaelis-Menten equation:

$$\frac{1}{V} = \frac{K_m(1)}{V_{max}[S]} + \frac{1}{V_{max}}$$

Plots of $1/V$ versus $1/[S]$ yield straight lines having a slope of K_m/V_{max} and an intercept on the ordinate at $1/V_{max}$ (Figure 3-2).

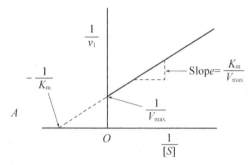

Figure 3-2 Lineweaver-Burk Plot

The Lineweaver-Burk transformation of the Michaelis-Menton equation is useful in the analysis of enzyme inhibition. Since most clinical drug therapy is based on inhibiting the activity of enzymes, analysis of enzyme reactions using the tools described above has been fundamental to the modern design of pharmaceuticals.

In this experiment, we willdetermine the activity of alkaline phosphatase (AKP) under different reaction conditions (substrate concentration) and calculate the value of K_m according to Lineweaver-Burk double-reciprocal plot.

AKP are the phosphate hydrolases that have a maximum reactivity at a relatively high pH (>7.0). AKP widely exist in both eukaryotic and prokaryotic cells. Some different substrates can be catalyzed by AKP with different K_m. In this experiment, a substrate called disodium phenylphosphate is used to measure theactivity of AKP. Disodium phenylphosphate can be hydrolyzed by AKP and produce phenol and phosphates(Figure 3-3).

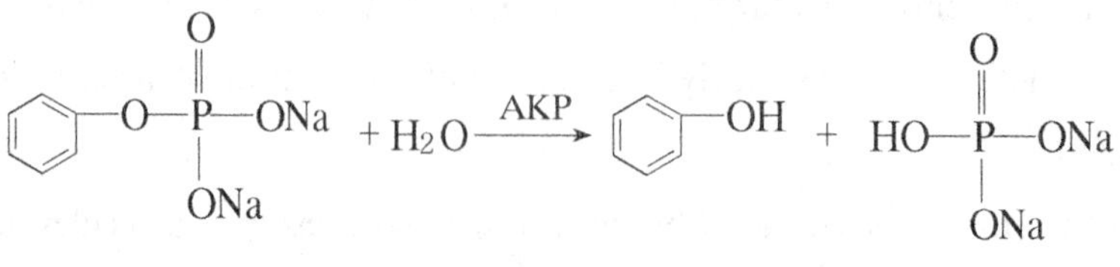

Figure 3-3 The reaction catalyied by AKP

The higher activity of AKP the more phenol is produced. So the concentration of phenol varies in proportion with the activity of AKP. Phenol and 4-aminoantipyrine can be oxidized to quinone derivatives(Figure 3-4). The more phenol active as substrate, the more quinone derivatives are produced, which are red compounds with maximumabsorption peak at 510 nm.

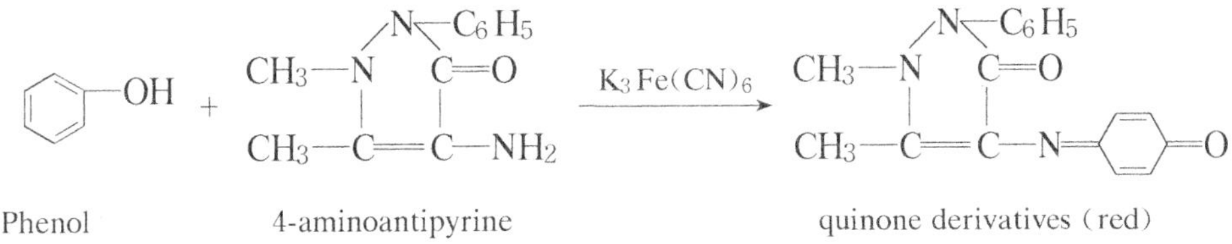

Figure 3-4 The colorimetric reaction of phenol

3.3 Procedures

3.3.1 The impact of substrate's concentration on velocity of enzyme reaction

(1) Take 6 test tubes and label them as 1~6(Table 3-1).

Table 3-1 K_m value determination of AKP

Reagent (mL)	Tube number					
	1	2	3	4	5	6
0.01 mol/L substrate solution	0.10	0.20	0.30	0.40	0.80	0.0
pH10.0, 0.1 mol/L carbonate buffer	0.70	0.70	0.70	0.70	0.70	0.70
ddH$_2$O	1.10	1.00	0.90	0.80	0.40	1.20
Mix well and incubate at 37℃ water bath for 5 min						
AKP solution	0.10	0.10	0.10	0.10	0.10	0.10
Final concentration of substrate (mmol/L)	0.5	1.0	1.5	2.0	4.0	0.0

(2) Mix well and incubate at 37℃ for 15 min.

(3) Add 1.1 mL of alkaline solution into each tube to terminate reaction.

(4) Add 1.0 mL of 0.3% 4-amino-antipyrine(AAP) and 2.0 mL of 0.5% potassium ferricyanide into each tube, mix well then place them for 10 minutes at room temperature (RT).

(5) #6 tube is used as blank. Measure the A510 of eachsample and the blank tube is used for the zero setting. Plot 1/A510 against 1/[S] and calculate K_m.

3.3.2 Plot a Standard Curve of Phenol Content

(1) Take 6 test tubes and label them as 1~6, #1 tube is used as blank(Table 3-2).

(2) Mix well and incubate at RT for 15 minutes. Measure the A510 of samples and the blank tube is used for the zero setting.

(3) A510 is plotted against phenolconcentration.

Table 3-2 Standard curve of AKP

Reagent (mL)	Tube number					
	1	2	3	4	5	6
0.1 mg/mL phenol standard solution	0.0	0.05	0.10	0.20	0.30	0.40
ddH$_2$O	2.0	1.95	1.90	1.80	1.70	1.60
Mix well and incubate at 37℃ water bath for 5 min						
Alkaline solution	1.10	1.10	1.10	1.10	1.10	1.10
0.3% 4-amino antipyrine	1.0	1.0	1.0	1.0	1.0	1.0
0.5% potassium ferricyanide	2.0	2.0	2.0	2.0	2.0	2.0

Mix well then place them for 10 minutes at room temperature(RT).

Notes:

① Pipetting the sample should be accurate and correct.

② Standard curve must be a straight line across base point (0, 0).

3.3.3 Results and Analysis

(1) Calculate K_m value of AKP by plotting Lineweaver-Burk line according to the data from table 1.

(2) One unit of enzyme activity is defined by the amount of enzyme that catalyzes the formation of one milligram of product (phenol) in 15 minutes at 37℃.

(3) Calculate the enzyme activity according to the standard curve of phenol content.

3.3.4 Reagents

(1) 0.01 mol/L substrate solution: weighing disodium phenyl phosphate

($C_6H_5PO_4Na_2 \cdot 2H_2O$) 1.016 g or disodium phenyl phosphate ($C_6H_5PO_4Na_2$) 0.872 g. Dissolve it with boiled distilled water, add 1,000 mL water and 4 mL antiseptic agent-chloroform. Store the solution in a brown bottle at 4℃ for 1 week.

(2) 0.1 mol/L carbonate buffer (pH = 10): weighing 6.36 g of anhydrous sodium carbonatedissolved in distilled water and add water to 1 000 mL.

(3) 0.01 mol/L KH_2PO_4: weighing 0.136 g of KH_2PO_4 and completely dissolved in 100 mL of distilled water.

(4) Alkine solution: measuring 20 mL 0.5 mol/L NaOH and 20 mL 0.5 mol/L $NaHCO_3$, mix well and then add water to 100 mL.

(5) 0.5% potassium ferricyanide: weighing 5 g of Potassium ferricyanideand 15 g of boric acid dissolved in 400 mL of water, mix well and add water to 1 000 mL and store in brown bottle in dark.

(6) 0.3% 4-aminoantipyrine: weighing 0.3 g of 4-aminoantipyrine and 4.2 g of $NaHCO_3$ dissolved in water and add water to 100 mL, stored in brown bottle at 4℃.

(7) Preparation of standard solution of phenol(1.0 mL≈0.10 mg).

① Weighing crystalline phenol 1.50 g, dissolved in 0.1 mol/L HCl, set the volume to 1 L as stock solution.

② Calibration: Take 25 mL of the stock solution, add 0.1 mol/L NaOH 55 mL, heated to 65 ℃, then 25 mL of 0.1 mol/L iodine solution added to cover for 30 min, add concentratedhydrochloric acid (HCl) 15 mL, 0.1% starch as indicator, with 0.1 mol/L sodium thiosulfate titration, the reaction is as follows: according to the reaction 3 molecule of iodine (MW = 254) with one molecule of phenol (MW = 94) per mL of 0.1 mol/L of iodine solution (iodine 12.7 mg) equivalent to the number of milligrams of the phenol was $12.7 \times 9.4/3 \times 254 = 1.567$.

Sodium thiosulfate in 25 mL of iodine solution by x mL, then 25 mL of phenol solution of phenol was 1.567 mg (25 − x).

③ according to the calibration results, stock solution is diluted with distilled water to 0.1 mg/mL as a standard solution when used.

3.3.5 Instruments

(1) Water bath.

(2) Spectrophotometer.

(3) Glass test tube.

(4) Graduated pipette.

3.3.6 Thought Questions

(1) What is the significance of measuring K_m?

(2) Compare the differences among three reversible inhibitions.

Experiment 4

Centrifugation Technique — Adaptive Immune Response: Isolation and Identification of Lymphocyte

4.1 Principles

4.1.1 Principles of Centrifugation

Centrifugation is a widely used conventional approach to isolate specific fraction from mixture based on the sedimentation of particles in suspension by centrifugal force. F is termed as sinking force, which acts on the sedimented particles in liquid under the gravity.

$$F = \text{Gravity-Buoyancy} = V\rho g - V\delta g = Vg(\rho - \delta)$$

V: volume of particle; ρ: density of particle; δ: density of media; g: acceleration of gravity. If $\rho > \delta$, then F denote negative buoyancy.

During centrifugation, particles in suspension move in a circle with r radius. Centripetal force acted on particle is called $V\delta r\omega^2$. ω denotes the angular velocity of circular motion. The centripetal force maintains uniform circular movement of particle with radius r is $V\rho r\omega^2$. If $\rho \neq \delta$, the difference between the actual centripetal force and the centripetal force maintains uniform circular motion is the value of $Vr\omega^2(\rho - \delta)$.

when $\rho > \delta$, the actual centripetal force acting on particles can not maintain its circular movement with radius r, the particles move in the direction away from the center of the circle. The larger value of ρ minus δ, the faster it escapes.

By comparing the above mathematical equation, $r\omega^2$ is equivalent to g, so increasing the rotating speed of centrifuges, the value of $r\omega^2$ is far more than the value of g, which consequently increases the sedimentation velocity of the particle during centrifugation.

Centrifugal force make the particle break away from circular motion, the size of the centrifugal force determines the separation effect. The relative centrifugal force(RCF) can be expressed by the following equation:

$$F = \frac{4\pi^2 r (60 \text{ r/min})^2}{g}$$

g: Acceleration of gravity ($980.6 \text{ cm}^2/\text{s}^2$);

r/min: Revolutions per minute;

r: The centrifugal radius.

The centrifugal force in g unit can also be expressed as:

$$F(g) = r \cdot (\text{r/min})^2 \times 1.118 \times 10^{-5}$$

4.1.2 The Classification of Centrifugation

Based on their spinning speed, centrifugation technique can be divided into 3 types: If the spinning speed is less than 6 000 r/min, it is called low speed centrifugation. If the centrifugation speed is between 6 000 r/min and 25 000 r/min, it is called high-speed centrifuge. If it is higher than 30 000 r/min, it is called ultracentrifugation. Depending on the different uses of the centrifugation, centrifugation can also be divided into analytical and Preparative centrifuge. The preparative centrifugation technique is further subdivided into fraction centrifugation technique and density gradient centrifugation. Especially, ultracentrifugation technique is used for isolation of biomacromolecule.

Various kinds of particles in an inhomogenous suspension precipitate with different velocity. The distal particles far away from axis center precipitate the fastest, generally they are the largest particles. In order to separate the fraction of the inhomogeneity, different centrifugation speedcan be set from low to high speed sequentially or centrifuging at high-speed and low-speed alternatively, This separation method is relatively coarse, recovery rate is not high.

Density gradient centrifugation is a commonly used technique, which can simultaneously separate single or multiple fractions from mixtures with high resolution. Samples was added into the density gradient medium and centrifugation is performed. At equilibrium, the fraction with same density of centrifugation medium will retention there. Density gradient centrifugation can further be divided into rate-zonal and isopycnic centrifugation.

4.1.2.1 Rate-zonal centrifugation

In rate-zonal centrifugation, the sample is layered as a narrow zone on top of a density gradient (Figure 4-1). In this way the faster sedimenting particles are not contaminated by the slower particles as occurs in differential centrifugation. The gradient stabilizes the bands and provides a medium of increasing density and viscosity.

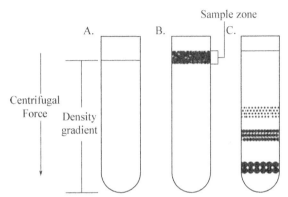

Sample is layered as a narrow zone on the top of a density gradient (B). Under centrifugal force, particles move at different rates depending on their mass (C)

Figure 4-1 Rate-Zonal Centrifugation

The speed at which particles sedimentation depends primarily on their size and mass instead of density. As the particles in the band move down through the density medium, zones containing particles of similar size form as the faster sedimenting particles move ahead of the slower ones. Because the density of the particles is greater than the density of the gradient, all the particles will eventually form a pellet if centrifugation is long enough.

4.1.2.2 Isopycnic separation

It is also called buoyant or equilibrium separation, particles are separated solely on the basis of their density. Particle size only affects the rate at which particles move until their density is the same as the surrounding gradient medium (Figure 4-2). The density of the gradient medium must be greater than the density of the particles to be separated. By this method, the particles will never sediment to the bottom of the tube, no matter how long the centrifugation time. The common media of isopycnic centrifugation is alkali metal salt solution, such as cesium salt or rubidium salts.

Ultracentrifugation is used in separation of subcellular structure and the preparation of biomacromolecule. The current speed can reach more than 85 000 rpm/min. Under the action of overspeed centrifugal field, centrifugal force is greater than the molecular diffusion

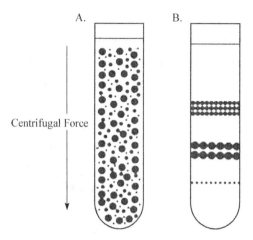

Starting with a uniform mixture of sample and density gradient (A) under centrifugal force, particles move until their density is the same as the surrounding medium (B).

Figure 4-2　Isopycnic Centrifugation

force, biological macromolecules settlement gradually. Due to different molecular weight and different molecular shape, their settling velocity are different, so they are separated.

The settling velocity of Biological macromolecules under the action of per unit centrifugal force field is called sedimentation coefficient. By convention, all sedimentation coefficients are expressed in the Svedberg units. One Svedberg (S) unit is 10^{-13} second. Sedimentation coefficient is defined as the ratio of a particle's sedimentation velocity to the acceleration that is applied to it (causing the sedimentation). The sedimentation speed (in ms^{-1}) is also known as the terminal velocity.

$$s = \frac{v_t}{a}$$

For example, the sedimentation coefficient of immunoglobulin is 7, that is, 7×10^{-13} seconds.

In this experiment, density gradient centrifugation is used to separate lymphoid cells. lymphocytes including T cells and B cells, and are an important component in the adaptive immune response. In many cases, the number of lymphocyte and the function is an important index of the body's immune system function.

Experiment 4 (1) The Separation of Monocytes by Density Gradient Centrifugation

4(1).1 Objective

Appreciating the principle and method of lymphocytes separation and identification.

4(1).2 Principles

According to the principle of the particle sedimentation in physics, the particles distribute to the different position in their settlement movement for its different density. Bydensity gradient centrifugation with the medium bearing approximately the same density as the being separated cells, the target cells can be collected.

The separation medium commonly used to separat emononuclear cells is composed of Ficoll and Hypaque mixed in certain proportions. Its specific gravity is (1.077 ± 0.001) g/L at 20℃, the gravity of lymphocytes and monocytes is lower than that of the separation medium. It is about 1.070 g/L, but the gravity of the granulocytes and erythrocytes is the highest and is about 1.092 g/L. Following centrifugation, the cells distribute according to their specific gravity in the density gradient. Lymphocytes and monocytes are located upper levels of separation medium, while granulocytes and erythrocytes settled to the bottom of the tube. Thus, lymphocytes and monocytes are separated.

Requirements of Separation medium for living cell:

① Can produce density gradient, high density while low viscosity.

② pH neutral or easily adjusted to neutral.

③ High concentration while osmotic pressure is not large.

④ Non-toxic to the cell.

Commonly used arethe lymphocyte separation medium: ① Ficoll; ② Dextran.

4(1).3 Procedures

(1) The mice is sacrificed with cervical dislocation, open abdominal cavity, cut peritoneum, spleen was dissected.

(2) Spleen was placed on 100 mesh sieve in culture dishes, totally 8 mL of PBS was added and simultaneously spleen was grinded with a grinding rod. The grinding solution was filtered through the sieve, cells suspension from spleen was collected.

(3) The suspension was centrifuged in a horizontal centrifuge at 1 500 rpm/min for 5 minutes. The some supernatant was discarded and the cell pellet was resuspended with 4 mL PBS.

(4) 4 mL of the spleen cells suspension (obtained from procedure3) was transferred to a new test tube containing 3 mL lymphocyte separation medium(Ficoll, density 1 088).

Note:
The cells suspension was carefully added so that the surface of the separation medium is not disturbed. The volume ratio of cells suspension and lymphocyte separation medium is 4 : 3. Mixture obtained from step 4 was centrifuged in a horizontal centrifuge at 1 800 rpm/min for 30 minutes at 20℃ (rate of rise and fall of the centrifuge will be set to 2).

(5) The test tube was taken out carefully. From bottom to top of the tube, it can be seen red blood cells, separation fluid and PBS, the faint white cell layer between PBS, which is the separated mononuclear cells (Figure 4-3).

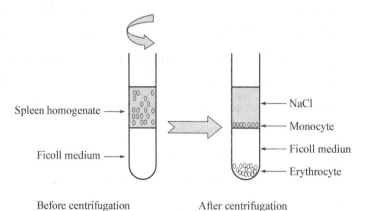

Figure 4-3 **The lymphocytes separated after Ficoll-Hypaque density gradient centrifugation**

(6) The layer of mononuclear cells from the interface between the PBS and lymphocyte separation medium was aspirated with pipettor carefully and then transferred into fresh 2 mL of PBS in another test tube. It was centrifuged at 1 500 rpm/min for 5 min. Finally the supernatant was discarded and the cell pellet was washed again.

(7) The pellet was resuspended in with $0.05\sim0.1$ mL PBS, a drop (10 μL) was taken to a adhesion slide, after natural dry, cell types can be identified.

Notes:

Spleen cells suspension should be add to the test tube along the tube wall slowly, make spleen cells mixed suspension and separation medium form the obvious interface, maintain complete interface, avoid to disturb interface to affect the separation efficiency.

1) Reagents

(1) Kunming mice, Ficoll lymphocyte separation medium (1.088), PBS.

(2) Dissecting instrument, 100 mesh sieve, sterile Petri dishes, centrifuge tube.

(3) The bench centrifuge machine, microscope, cell count plate, cell counter.

(4) Pointed pipettors, rubber suction head, slides, blotting paper, etc.

2) Thought Questions

(1) What are the key steps of this experiment?

(2) From which interface can we get more lymphocytes? Are there other methods to separate blood mononuclear cell?

(3) If you want to further analyze the mononuclear cells in cell types such as B cells, what kind of methods do you want to take?

Experiment 4 (2) Lymphocyte Identification
——Counting the Number of B Lymphocyte

4(2).1 Principles

SmIgM(Surface membrane immunoglobulin M) is a B cell antigen, recognition receptor, and is also a B cell specific surface marker, a number of specific methods can be used for its detection(Figure 4-4). In this experiment, mIgM is detected by immunohistochemistry. Horse radish peroxidase labeled anti-IgM antibody was incubated with lymphocytes, the labelled IgM antibody can bind to B cell surface IgM, through the DAB chromogenic method, specific staining will appear on the cell membrane observed under ordinary optical microscope and B lymphocytes can be identified by this method.

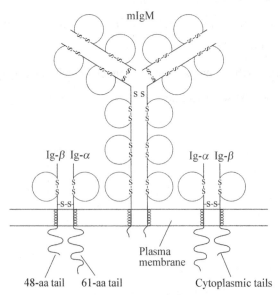

Figure 4-4 Specific Surface Markers of B Lymphocyte

4(2).2　Procedures

(1) The separation of lymphocytes as mentioned in the Experiment 4 (1).

(2) Fixation: at room temperature, using methanol or 10% formalin to fix cells, after natural drying, mark sample boundary with crayons, wash three times using distilled water. Draw two area on one slide, a positive area, a negative control area (without antibody).

(3) Blocking: prepare blocking solution (6 μL hydrogen peroxide with 1 mL PBS containing 10% rabbit serum). Block non-specific binding sites with blocking fluid for 30 min at room temperature, wash three times with PBS.

(4) Antibody incubation: incubate with peroxidase labeled rabbit anti mouse IgM antibody diluted by 1 : 100 in wet box, for 30 min, in 37℃ constant temperature box (or at room temperature, for 1 h), wash three times in PBS.

(5) Visualizaiton: add DAB buffer, develop for 20 min at room temperature. Wash three times using distilled water. Embed in resin with coverslip and observe with microscope.

(6) Nuclear Fast Red staining: Nuclear Fast Red staining for 10 min, washed in distilled water. (optional)

(7) Dehydrate through gradient ethanol containing 85% alcohol 20 s; 95% alcohol 1 min, 2 changes of 100% absolute alcohol, 2 minutes each. Clear in 2 changes of xylene, 2 min each. Mount with Neutral balsam based mounting medium. (optional)

1) Results and Analysis

Under the microscope, the cells with annular or mottled purplish blue staining on membrane are positive cells. The nucleus was stained by Nuclear Fast Red. In the same field, count positive lymphocytes and the total number of lymphocytes (100 lymphocytes), calculate the percentage of positive B cells.

Notes:
① Before adding DAB solution, unbound primary antibody should be washed away completely to avoid false positive results.
② Both positive and negative control are needed to label.

2) Reagents

(1) Kunming mice, Ficoll (1.088) and lymphocyte separation medium, physiological

saline, PBS, and distilled water.

(2) 10% formalin, dilution blocking fluid (containing 0.6% H_2O_2, 10% rabbit serum in PBS).

(3) The peroxidase labeled rabbit anti mouse IgM antibody, DAB chromogenic fluid (purple blue).

(4) Nuclear Fast Red solution: Nuclear Fast Red 0.1 g, 5 g aluminum sulfate, distilled water 100 mL, thymol 50 mg.

3) Instruments

(1) Dissecting instrument, 100 mesh sieve, sterile petri dishes, centrifuge tube.

(2) The bench centrifuge, microscope, cell count plate, cell counter.

(3) Pointed pipettes, rubber suction head, slides, blotting paper, etc.

4) Thought Questions

(1) Why do we incubated blocking buffer before adding primary antibody?

(2) What are the key steps of this experiment? What are the possible problem if we failed to get the positive cells?

Experiment 5

Polyacrylamide Gel Electrophoresis of Proteins

5.1 Objectives

(1) Understand the principles of protein separation by SDS polyacrylamide gel electrophoresis (SDS-PAGE).

(2) Be able to use SDS-PAGE technique to separate proteins.

5.2 Principles

SDS-PAGE is the most widely used method for qualitatively analyzing protein mixtures. It is particularly useful for monitoring protein purification, and because the mechanism for separation of proteins in this method is according to protein size, it can also be used to determine the relative molecular mass of proteins.

5.2.1 Formation of Polyacrylamide Gels

Crosslinked polyacrylamide gels are formed from the polymerization of acrylamide monomer in the presence of smaller amounts of N,N'-methylene-bis-acrylamide (normally referred to as "bis-acrylamide") (Figure 5-1). Note that bis-acrylamide is composed of two acrylamide molecules linked by a methylene group and is used as a crosslinking agent. Acrylamide monomer is polymerized in a head-to-tail fashion into long chains, and occasionally a bis-acrylamide molecule is built into the growing chain, thus introducing a second site for chain extension. Proceeding in this way, a crosslinked matrix of fairly well-defined structure is formed. The polymerization of acrylamide is an example of free-radical catalysis, and is initiated by the addition of ammonium persulfate and the base N,N,N',N'-

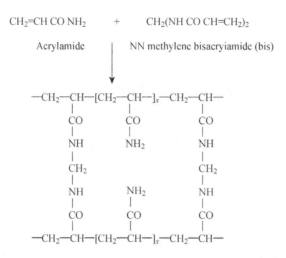

Figure 5-1 Polymerization of acrylamides

tetramethylenediamine (TEMED). TEMED catalyzes the decomposition of the persulfate ion to give a free radical (i.e., a molecule with an unpaired electron):

$$S_2O_8^{2-} + e^- \longrightarrow SO_4^{2-} + SO_4^-$$

If this free radical is represented as R· (where the dot represents an unpaired electron) and M as an acrylamide monomer molecule, then the polymerization can be represented as follows:

$$R\cdot + M \longrightarrow RM\cdot$$
$$RM\cdot + M \longrightarrow RMM\cdot$$
$$RMM\cdot + M \longrightarrow RMMM\cdot, \text{etc.}$$

In this way, long chains of acrylamide are built up, being crosslinked by the introduction of the occasionalbis-acrylamide molecule into the growing chain.

5.2.2 The Use of Stacking Gels

For both SDS and buffer gels, samples may be applied directly to the top of the gel in which protein separation is to occur (the separating gel). However, in these cases, the sharpness of the protein bands produced in the gel is limited by the size (volume) of the sample applied to the gel. Basically the separated bands will be as broad (or broader, owing to diffusion) as the sample band applied to the gel. For some work, this may be acceptable, but most researchers require better resolution than this.

This can be achieved by polymerizing a short stacking gel on top of the separating gel. The purpose of this stacking gel is to concentrate the protein sample into a sharp band before it enters the main separating gel, thus giving sharper protein bands in the separating gel. This modification allows relatively large sample volumes to be applied to the gel without any loss of resolution. The stacking gel has a very large pore size (4% or 5% acrylamide) which allows the proteins to move freely and concentrate, or stack under the effect of the electric field.

Sample concentration is produced by isotachophoresis of the sample in the stacking gel. The band-sharpening effect (isotachophoresis) relies on the fact that the negatively charged glycinate ions (in the reservoir buffer) have a lower electrophoretic mobility than the protein-SDS complexes, which in turn, have lower mobility than the Cl^- ions if they are in a region of higher field strength. Field strength is inversely proportional to conductivity, which is proportional to concentration. The result is that the three species of interest adjust their concentrations so that $[Cl^-] > [\text{protein-SDS}] > [\text{glycine}]$. There are only a small quantity of protein - SDS complexes, so they concentrate in a very tight band between the glycinate and Cl^- ion boundaries. Once the glycinate reaches the separating gel, it becomes more fully ionized in the higher pH environment and its mobility increases (The pH of the stacking gel is 6.8 and that of the separating gel is 8.8). Thus, the interface between glycinate and the Cl^- ions leaves behind the protein - SDS complexes, which are leftto electrophorese at their own rates.

5.2.3 SDS-PAGE

Samples to be run on SDS-PAGE are first boiled for 5 min in sample buffer containing a reducing reagent, such as α-mercaptoethanol or dithiothreitol (DTT), and SDS. Theα-mercaptoethanol or DTT reduces any disulfide bridges present that are holding together the protein tertiary structure. SDS ($CH_3-[CH_2]_{10}-CH_2OSO_3-Na^+$) is an anionic detergent and binds strongly to, and denatures, the protein. Each protein in the mixture is therefore fully denatured by this treatment and opens up into a rod-shaped structure with a series of negatively charged SDS molecules along the polypeptide chain. On average, one SDS molecule binds for every two amino acid residues. The original native charge on the molecule is therefore completely swamped by the SDS molecules. The sample buffer also contains an ionizable tracking dye usually bromophenol blue (BPB) that allows the electrophoretic run to be monitored, and glycerol which gives the sample solution density, thus allowing the sample to settle easily through the electrophoresis buffer to the bottom when loaded into the wells.

When the main separating gel has been poured between the glass plates and allowed to set, a shorter stacking gel is poured on top of the separating gel, and it is into this gel that the wells are formed and the proteins loaded. Once all the samples are loaded, a voltage is passed through the gel. Once the protein samples have passed through the stacking gel and have entered the separating gel, the negatively charged protein-SDS complexes continue to move toward the anode, and because they have the same charge per unit length, they travel into the separating gel under the applied electric field with the same mobility. However, as they pass through the separating gel the proteins separate, owing to the molecular sieving properties of the gel. Quite simply, the smaller the protein, the more easily it can pass through the pores of the gel, whereas large proteins are successively retarded by frictional resistance owing to the sieving effect of the gel.

Being a small molecule, the BPB dye is totally un-retarded and therefore indicates the electrophoresis front. When the dye reaches the bottom of the gel the voltage is turned off and the gel is removed from between the glass plates, shaken in an appropriate stain solution (usually Coomassie brilliant blue) for $0.5 \sim 1$ hour, and then washed in distaining solution several times. The destaining solution removes unbound background dye from the gel, leaving stained proteins visible as blue bands on a clear background. A typical mini-gels (e.g., Bio-Rad mini-gel) run at 200 V. Constant voltage can run in about 1 h, and require only $0.5 \sim 1$ h staining. Most bands can be seen within 1 h of destaining.

The vertical mini-gel system from Bio-Rad is used in this experiment (Figure 5-2).

5.3 Procedures

(1) Samples to be run are first denatured in sample buffer by heating to $95 \sim 100\,^\circ\text{C}$ for 5 min.

(2) Clean the internal surfaces of the gel plates with detergent, then dry, and join the gel plates together to form the cassette, and clamp it in a vertical position (Figure 5-3).

(3) Mix the following reagents in a small beaker to make the separating and stacking gel (Table 5-1 and Table 5-2):

Part 1 Biochemistry Experiment

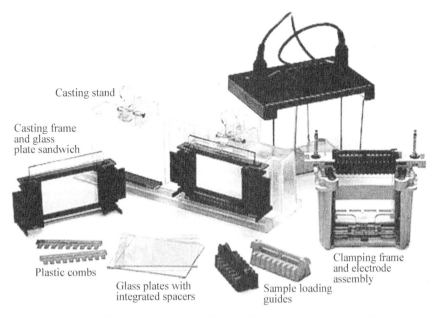

Figure 5-2　The entire apparatus of gel making and protein electrophoresis

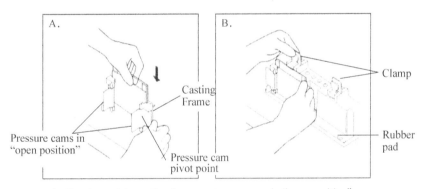

A. Put glasses into casting frame and press cams in "open position".
B. Put assembled casting frame in A to casting stand and fit with clamp.

Figure 5-3　Assembly of cassette for making gel

Table 5-1　Separating gel

Components(Separating)	Volume(mL)
H_2O	4.3
40% Arc/Bis	3.0
1.5M Tris-Cl (pH8.8)	2.5
10% SDS	0.1
10% APS	0.1
TEMED	0.005

Table 5-2　Stacking gel

Components(Stacking)	Volume(mL)
H_2O	2.37
40% Arc/Bis	0.5
0.5 M Tris-Cl (pH6.8)	1.0
10% SDS	0.04
10% APS	0.04
TEMED	0.005

(4) Gently shake the beaker to ensure even mixing. The addition of TEMED will initiate the polymerization reaction and although it will take about 20 min for the gel to set, the time can depend on environmental temperature.

(5) Use a pipette to transfer this separating gel mixture to the gel cassette by running the solution carefully down one edge between the glass plates. Continue to add the solutionuntil it reaches a position about 1 cm from the bottom of the comb that will form the loading wells.

(6) To ensure that the gel sets with a smooth surface, very carefullyrun distilled water down one edge into the cassette using a pipette. Because of the great difference in density between the water and the gel solution, the water will spread across the surface of the gel without serious mixing. Continue adding water until a layer of about 2 mm exists on top of the gel solution.

(7) The gel can now be left to set. When set, a very clear refractive index change can be seen between the polymerized gel and overlaying water.

(8) When the separating gel has set, pour off the overlaying water. Add the stacking gel solution to the gel cassette until the solution reaches the cutaway edge of the gel plate. Place the well-forming comb into this solution, and leave to set. This will take another 20 min. Refractive index changes around the comb indicate that the gel has set.

(9) Carefully remove the comb from the stacking gel, and then rinse out any nonpolymerized acrylamide solution from the wells using electrophoresis buffer. Remove any spacer from the bottom of the gel cassette, and assemble the cassette in the electrophoresis tank (Figure 5-4). Fill the top reservoir with electrophoresis buffer, and look for any leaks from the top tank. If there are no leaks fill the bottom tank with electrophoresis buffer.

(10) Samples now can be loaded onto the gel. Place a pipette tip through the buffer and locate it just above the bottom of the well. Slowly deliver the sample into the well. 10 μL samples are appropriate for mini-gels. The dense sample buffer ensures that the sample settles to the bottom of the loading well. Follow the same way to fill all the wells with samples or protein markers, and record the samples loaded.

(11) Connect the power pack to the apparatus, and pass a voltage of 200 V (constant voltage). Ensure your electrodes have correct polarity: all proteins will travel to the anode (+). In the first few minutes, the samples will be seen to be concentrated as a sharp band as it moves through the stacking gel. (Actually it is observing BPB not the protein, but of course the protein is stacking in the same way.) Continue electrophoresis until the BPB

Part 1 Biochemistry Experiment

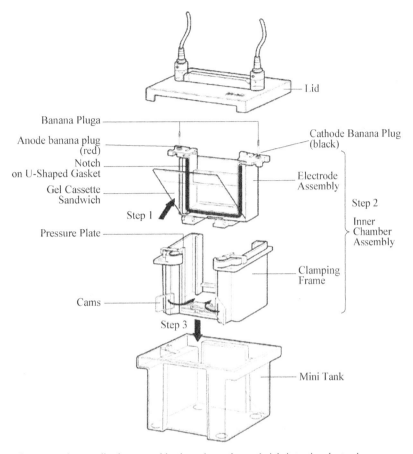

Step 1: as the arrow shown, firstly, assemble the gel casstle sandwich into the electrode.
Step 2: secondly, put the assembled electrode into the clamping frame and put cams in "close position" and finish the assemble of inner chamber.
Step 3: put the assembled inner chamber into the mini tank, finally, cover the mini tank with the lid.

Figure 5-4 Assembly of the electrophoresis tank

reaches the bottom of the gel. This will take about 1~1.5 h.

(12) Dismantle the gel apparatus, pry open the gel plates, remove the gel, discard the stacking gel, and place the separating gel in stain solution.

(13) Staining should be carried out with shaking for 30 min and destaining for 30~60 min(overnight is better). When the stain is replaced with destain, stronger bands will be immediately apparent, and weaker bands will appear as the gel destains.

Notes:

① Acrylamide is a potential neurotoxin and should be treated with great care. Its effects are cumulative, and therefore, regular users are at greatest risk. In particular, take care and avoiding direct contact with the solution.

② Fresh-made 10% Ammonium persulfate is best for making gels. But this solution may be also aliquoted and stored at −20℃ for later use.

③ Because the time for the polymerization of acrylamides depends on the room temperature, kept it over 20℃.

1) Results and analysis

Take pictures of stained gel for observation of protein bands.

2) Reagents

(1) 5X electrophoresis buffer: Tris (15.1 g), glycine (94 g), and SDS (5.0 g), add water to 1 L. No pH adjustment is necessary.

(2) Tris-buffers: ①1.5MTris-HCl, pH 8.8. ②1MTris-HCl, pH 6.8.

(3) Stock acrylamide solution: 40% acrylamide, 1% bis-acrylamide. Filter with filter paper and store at 4℃.

(4) 10% SDS.

(5) TEMED.

(6) 10% Ammonium persulfate, fresh-made.

(7) Protein stain: 0.25% Coomassie brilliant blue R250 in 45% methanol, 10% glacial acetic acid. Dissolve the dye in the methanol and water componentfirst, and then add the acetic acid. Filter the final solution through filter paper and store in dark.

(8) Destain: 45% methanol, 10% glacial acetic acid.

(9) 10X sample buffer: 0.5MTris-HCl, pH 6.8; 10% SDS; 1M DTT; 1% BPB; 50% Glycerol.

3) Instrument

(1) Vertical gel casting system.

(2) Vertical gel running system and power supply.

(3) Small beakers.

(4) Adjustable pipettes and tips.

4) Thought Questions

(1) What are the principles of protein separation by SDS-PAGE?

(2) What are the precautions during gel making?

Experiment 6

Effect of Hormone on Blood Sugar and Lipoprotein of Normal and Diabetic rat

6.1 Objectives

(1) To understand the principle of determination of blood glucose and lipoprotein.

(2) To appreciate the change of carbohydrate and lipid metabolism of diabetic rat.

(3) To appreciate the effect of hormone on blood glucose and the clinical significance of blood lipoprotein.

6.2 Principles

Streptozocin (STZ) can be used to construct diabetic rat model. STZ is a cellular toxic compound. It can be relatively targeted to β cells and induce type 1 diabetes by causing damage to DNA. In this experiment, we observe the blood sugar and lipid in control and diabetic rat established by STZ.

Glucose oxidase(GOD) method is used extensively to determine blood glucose in clinic because of its high specificity, low cost and convenient operation. Glucose is oxidized by glucose oxidase to produce gluconate and H_2O_2. Hydrogen peroxide is then oxidatively coupled with 4-amino-antipyrene(4-APP) and phenol in the presence of peroxidase(POD) to yield a quinoeimine dye that is measured at 505 nm. The absorbance at 505 nm is proportional to glucose concentration in the sample. Similarly, the oxidation of triglyceride and cholesterol release H_2O_2 and have the similar color reaction with glucose.

Notes:

The normalfasting blood glucose concentration in an adult human is 3.89 ~

6.11 mmol/L.

Hypoglycemic and hyperglycemic hormones are two kinds of hormones to regulate blood glucose in mammals. Insulin is the unique hypoglycemic hormone to lower the blood glucose. While the hyperglycemic hormones include glucagon, glucocorticoid, adrenalin et al.

6.3 Procedures

6.3.1 Construction of diabetic rat model

Adult male rats (Sprague-Dawley), weighing about 200 g, are divided into control group and diabetic group. Diabetic group is treated with STZ by intraperitoneal injection at the dose of 60mg/kg weight mass. The control group is treated with the same volume of citric acid solution. After 24 h, 48 h and 72 h of injection, blood sugar is measured respectively. Those rats with blood sugar greater than 13.8 mmol/L are identified as diabetic rats. Rats are fasted 16h in advance of the experiment.

6.3.2 Tail bleeding

Students work in pair to deal with one rat. The tail of rat is firstly warmed by immersing into water of 45°C (the tail also can be spread with dimethylbenzene). Then, the tail is cleaned with a cotton swab moistened with 70% ethanol. Snip off the very tip of the rat's tail (5 mm to the end). The rat's tail should be gently squeezed, starting at the base of the tail and moving towards the site of the cut, to get a drop of blood from the incision site. The drops of blood should be carefully brought into the tubes and mark the sample with Pre. 200~300 μL blood is good enough. Then the cut is staunched by compressing.

6.3.3 Administering hormone injection

Mark the control rats and diabetic rats. Both groups of rats are further divided into two sub-groups marked with group 1 and group 2. Group 1 is injected with insulin, 0.75 IU/kg (60 U/kg), taking 0.2~0.3 mL of blood into tube after 30 min injection and mark with Ins. Group 2 is injected with adrenaline, 0.1% adrenaline 0.2 mg/kg, 30 min later, taking 0.2~0.3 mL of blood into tube, marked with Adr.

Separating theserum from blood cells by 3 000 rpm centrifuge for 5 min and taking the serum supernatant for following determination.

6.3.4 Blood glucose determination (GOD-POD method)

As shown in Table 6-1, prepare the tabes and reagents to determine blood glucose concentration.

Table 6-1 Blood glucose assay

Reagent (μL)	Blank	Pre	Ins	Adr	Standard
Serum	—	10	10	10	—
GlucoseStandard(5.05 mmol/L)	—	—	—	—	10
ddH$_2$O	10				
Glucose reagent	1 000	1 000	1 000	1 000	1 000

These tubes are vibrated gently to mix and incubated at 37℃ for 10 min. Calibrate the spectrophotometer at 505 nm against sample blank (Blank). Record absorbance value of each sample.

6.3.5 Triglyceride (colorimetric method)

As shown in Table 6-2, prepare the tubes and reagents to determine blood triglyceride concentration.

Table 6-2 Blood triglyceride assay

Reagent (μL)	Blank	Pre	Ins	Adr	Standard
Serum	—	10	10	10	—
Triglyceride standard (2.26 mmol/L)	—	—	—	—	10
ddH$_2$O	10				
Triglyceride reagent	1 000	1 000	1 000	1 000	1 000

These tubes are vibrated gently to mix and incubated at 37℃ for 10 min. Calibrate the spectrophotometer at 546 nm against sample blank (Blank). Record absorbance value of each sample.

6.3.6 Total cholesterol (colorimetric method)

As shown in Table 6-3, prepare the tubes and reagents or determine blood cholesterol concentration.

Table 6-3 Blood cholesterol assay

Reagent(μL)	Blank	Pre	Ins	Adr	Standard
Serum	—	10	10	10	—
Cholesterol standard (5.17 mmol/L)	—	—	—	—	10
ddH$_2$O	10				
Cholesterol reagent	1 000	1 000	1 000	1 000	1 000

These tubes are vibrated gently to mix and incubated at 37℃ for 10 min. Calibrate the spectrophotometer at 505 nm against sample blank (Blank). Record absorbance value of each sample.

Notes:

If the sample is from plasma, the syringe and tubes should be properly moistened with heparin before being used.

6.3.7 Calculation

(1) Blood sugar/lipid(mmol/L) = $\dfrac{A_{sample}}{A_{standard}}$ × Concentration of standard

(2) Collect the data from different groups and fill in the foloowing Table 6-4 and take on overview about the change of indexes between control and diabetic mellius rats.

All the data are summarized in table 6-4 for furher comparison and stafistic.

6.3.8 Supplementtary

(1) References value

Rat serum/plasma, Limosis:

① blood glucose: 2.64~5.26 mmol/L;

② blood cholesterol: 1.0~1.5 mmol/L;

③ blood triglyceride: 0.4~0.7 mmol/L.

(2) Clinical significance:

① Blood glucose assay

a. Normal hyperglycemia: After food intake or in stress.

b. Abnormal hyperglycemia

- Diabetes mellitus: Insulin is absolutely and relatively insufficient.

Table 6-4 Summary sheet of data from total class

Treatment Index	Construct of DM rat		1 week after construction of DM*					
	Control	DM	Pre		Ins		Adr	
			Control	DM	Control	DM	Control	DM
Blood glucose (mmol/L)								
Blood triglyceride (mmol/L)								
Blood cholesterol (mmol/L)								

* Students establish DM rat at the end of last experiment, one week later at the normal time of next experiment, this experiment is further operated.

- Endocrine abnormal: Adrenal cortical hyperactivity (Cushing's syndrome), hyperthyroidism.
- Acromegaly.
- Obesity.
- Acute stress reaction, shock and convulsions

c. Normal hypoglycemia: starvation or after severe movement.

d. Abnormal hypoglycemia: Insulinoma, Adrenal cortical insufficiency Addison's disease, hypopituitarism.

② Blood lipoprotein assay

Indexes: Total Cholesterol (TC); Triglyceride (TG); LDL-Cholesterol (LDL-C); HDL-Cholesterol (HDL-C).

TC, TG and LDL-C are high risk factors of arteriosclerosis.

HDL-C is protective factor of arteriosclerosis.

6.4 Reagents

(1) Insulin.

(2) Adrenaline.

(3) Gelation inhibitor: heparin 12 500 U/2 mL.

(4) Blood glucose assay reagent (kit): glucose oxidase >10 U/mL, peroxidase>1 U/mL, phosphate 70 mmol/L, phenol: 5 mmol/L, 4-amino-antipyrene: 0.4 mmol/L, pH 7.0.

(5) Total cholesterol assay reagent(kit): Pipes (piperazine-N,N'-bis(2-ethanesulfonic acid)) 35 mmol/L, cholesterol oxidase>0.1 U/mL, phenol 28 mmol/L, sodium cholate 0.5 mmol/L, 4 - AAP 0.5 mmol/L, cholesterol esterase >0.2 U/mL, peroxidase > 0.8 U/mL pH 7.0.

(6) Triglyceride assay reagent (kit): Pipes 45 mmol/L, magnesium chloride 5.0 mmol/L, glycerol kinase >1.5 U/mL, lipoprotein lipase >100 U/mL, 3 - phosphate glycerol oxidase >4 U/mL, EHSPT (N-Ethyl-N-(2-hydroxy-3-sulfopropyl)-3-methylaniline) 3.0 mmol/L, 4 - amino antipyrene 0.75 mmol/L, peroxidase > 0.8 U/mL, ATP 0.9 mmol/L, pH 7.5.

(7) 0.05 mol/L citric acid solution: 0.1L×0.05 mol/L×210.14(molecular weight of citric aicd) =1.0507 g citric acid is dissolved in 100 mL ddH$_2$O and pH is adjusted to 4.5. Making a filtration sterilization by 0.22 μm filter paper.

(8) STZ is purchased from sigma.

6.5 Instruments

(1) Centrifuge.

(2) Pipette.

(3) 96-well plate.

(4) Eppendorf tube.

(5) Microplate reader.

6.6 Thought Questions

(1) How does insulin regulate blood glucose?
(2) Describe the classifications and functions of blood lipoproteins.

Experiment 7

Polymorphism Analysis of ACE Gene by Extracting Genome DNA of Buccal Epithelial Cells

7.1 Objectives

Learn to extract genome DNA from a small amount of tissue cells, and understand how PCR method is used to analyze gene polymorphism.

7.2 Principles

ACE (angiotensin converting enzyme, ACE) gene has 2 common variants of intron 16 characterized by the insertion (I) or deletion (D) of Alu element, which results in the three polymorphisms: type DD, ID and II. In this experiment, genome DNA of each students is extracted by alkaline lysis method, and then PCR is performed by particular primers of ACE gene binding to the upstream and downstream of Alu element insertion or deletion site. Polymorphism can finally be determined by length of PCR products. The length of PCR product of type II is 490 bp, type DD is 190 bp and type ID has both 490 bp and 190 bp.

7.3 Procedures

7.3.1 Extraction of genome DNA

(1) Use solution I 10 mL to rinse mouth for 20 s, collecting mouthwash.

(2) Mouthwash was centrifuged in 3 000 g×5 min in room temperature, then discarding

the supernatant.

(3) Add solution Ⅱ 250 μL to resuspend participation and transfer to 1.5 mL EP tube, then perform centrifuge in 3 kg×1 min, discarding supernatant.

(4) Add solution Ⅲ 250 μL to resuspend participation, vortex for 10 s (**Alkaline lysis step**)

(5) Heat at 99℃ ×5 min.

(6) Add solution IV 50 μL, vortex for 5 s (**Neutralization step**)

(7) Perform centrifuge in 3 kg×5 min.

(8) Transfer supernatant to a new 1.5 mL EP tube, and then take 5 μL supernatant as template for PCR.

7.3.2 PCR

(1) Add the following ingredients in a sterile 0.2 mL EP tube, and then mix.

2×PCR mix	10 μL
DNA template	5 μL
Primer mixture (10 μM each)	2 μL
ddH$_2$O	3 μL
Total	20 μL

(2) Perform the following PCR program:

94℃ for 5 min, and then entered 35 cycles of 94℃ for 30 s, 55℃ for 30 s, 72℃ for 40 s, after cylces, one more extension of 72℃ for 10 min, finally PCR products were put in 4℃ for later detection or −20℃ for storage.

7.3.3 Electrophoresis

Casting of 1.5% agarose gel: 0.75 g agarose powder was dissovled in 50 mL 1× TBE electrophoresis buffer, and was heated in microwave oven for 3 min, after cool to 60℃, add 2.5 μL ethidium bromide (EB) dye and mixed immediately, then pour the melting gel to tank. A comb was placed in the tank to create wells for loading sample. When the gel was set, remove the comb and load 20 μL of PCR product to gel well. Electrophoresis was performed under 5 volt/cm.

Notes:

① Rinsing mouth fully for 20 s.

② Vortex gently and quickly to avoid breaking DNA after cell lysis.

③ EB is an intercalating agent. When EB is under ultraviolet, it will fluoresce with an orange color and the color will be intensified 20 fold after EB binds to DNA. EB may be a mutagen and its use should be careful.

④ Concentration ratio between template DNA and primer is important in PCR reaction. Primers are sequence specific fragments designed according to target gene, and templates are DNA mixture. In order to increase specificity of amplification, excessive amount of primers vs low concentration of templates is required. But high concentration of primers may produce dimmers and non-specific amplification, and low concentration of templates may decrease efficiency of amplification, or produce no products. Usually, the concentration of primer mixtures is 1pmol/μL, and concentration of templates can be determined by preliminary experiments.

⑤ In primer designing, the upstream and downstream primers should have similar Tm. If there is a big difference of Tm between upstream and downstream primers, annealing temperature is set according to the lower Tm of the primer.

7.4 Results and Analysis

(1) As shown in Figure 7-1, three polymorphisms of ACE can be identified by electrophoresis.

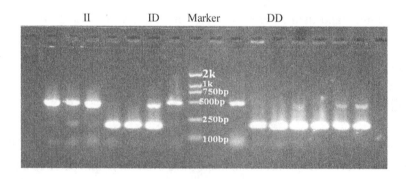

II: Homogenous insertion type
DD: Homogenous deletion type
ID: Heterogenous type of insertion and deletion

Figure 7-1 Electrophoresis result of ACE gene amplification by PCR

(2) Clinical Significance

ACE level was shown to be high in type DD, middle in ID and low in II. Patients with left vetricular hypertrophy (LVH) have high frequency of type DD.

7.5 Reagents

(1) Genome DNA Extraction

① Solution I: 4% sucrose:4 g is dissolved in 100 mL ddH$_2$O.

② Solution II: 10 mM NaCl and 10 mM EDTA (pH7.5).

③ Solution III: 50 mM NaOH.

④ Solution IV: 1.0 M Tris-HCl (pH 7.5).

(2) PCR kit

① dNTP Mix(10mM/each).

② Taq DNA polymerase.

③ 10×Buffer [200 mM Tris-HCl(pH 8.4),500 mM KCl, 20 mM MgCl$_2$].

(3) ACE Primers

Forward:5'-CTGGAGACCACTCCCATCCTTTCT-3'.

Reverse:5'-GATGTGGCCATCACATTCGTCAGAT-3'.

(4) Electrophoresis

1×TBE: Tris 54 g, boric acid 27.5 g and 0.5 M EDTA 20 mL are dissolved in 5 L H$_2$O.

7.6 Instruments

(1) Centrifuge.

(2) PCR thermocycle instrument.

(3) Pipettes.

7.7 Thought Questions

(1) What is the theoretical basis and method of gene diagnosis?

(2) Can you show the clinical significance of gene polymorphism?

Part 2
Cell Biology Experiment

Experiment 8

Optical Microscope and Cell Cycle

8.1 Objectives

(1) Be familiar with the major components of optical microscope.

(2) Be able to use optical microscope correctly.

(3) Grasp the basic method of biological drawing.

(4) Learn to make a temporary mount.

(5) Learn how to measurethe size of a cell under microscope.

(6) Master the distinctive features of different stages of mitosis.

8.2 Principles

8.2.1 Optical Microscope

The opticalmicroscope, alsoreferred to as the "light microscope", is a very important and useful instrument in cell biology. It uses visible light and a system of lenses to magnify images of small specimens. With the help of an optical microscope, we can observe cells and subcellular structures which we cannot see with our naked eyes.

8.2.1.1 Components of an Optical Microscope

All modern optical microscopes designed for viewing samples by transmitted light share the same basic components. They consist of two systems, one is mechanical system, and the other is optical system. The mechanical system usually includes a base, a stage, a revolving nosepiece and focus wheels, and the optical system is composed of a condenser, objective lens and ocular lens(Figure 8-1).

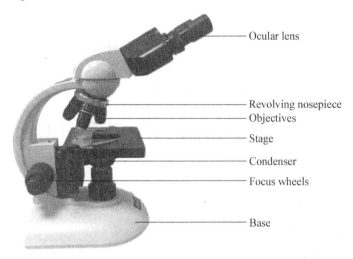

Figure 8-1　Major components of a light microscope

(1) Mechanical system

① Base: The base is the lowest part of an optical microscope, and itlets the microscope rest on the desk.

② Stage: Stage is a platform which supports the specimen to be observed.

③ Revolving nosepiece: Revolving nosepiece is the part that hold saset of objectives. The objectives can be changed through rotating the revolving nosepiece.

④ Focus wheels: Focus wheels move the stage up and down with separate adjustment knobs for coarse and fine focusing.

(2) Optical system

① Condenser: The condenser is a lens designed to focus light from the illumination source onto thespecimen. The condenser may also include other features, such as a diaphragm and/or filters, to manage the quality and intensity of the illumination.

② Objectives: The objective is usually in a cylinder housing containing a glass single or multi-element compound lens. Typically there will be three objectives screwed into a revolving nosepiece which may be rotated to select arequired objective lens. These objectivesare designed to be parfocal, which means that when one changes from one objectiveto another on a microscope, the sample stays in focus. Microscope objectives are characterized by two parameters, magnification and numerical aperture. The former typically ranges from $5\times$ to $100\times$ while the latter ranges from 0.14 to 0.7, corresponding to focal lengths of about 40 mm to 2 mm, respectively. Objectives with higher magnifications normally have a higher numerical aperture and a shorter depth of field in the resulting image.

③ Ocular lens: The eyepiece, or ocular, is inserted into the top end of the body tube. Its function is to bring the image into focus for the eye. Typical magnification values for eyepieces 10×.

8.2.1.2 Magnification of an Optical Microscope

The visible light is focused on the specimen by the condenser, and then passes through objective lens and ocular lens. Each of the latter two lenses magnifies the image of the specimen. Therefore, the actual power or magnification of alight microscope is the product of the powers of the ocular (eyepiece) and the objective lens. The maximum normal magnifications of the ocular and objective are 10× and 100× respectively giving a final magnification of 1 000×.

There are three key factors, resolution, illumination and contrast, thatdecide what can be seen under a light microscope. Resolution is depend on the resolving power of an objective, which is the ability to distinguishtwo objects separately. The resolving power is in inverse proportion to the wavelength of substance, and in direct proportion to the numerical aperture. Using oil immersion lens will improve the resolution. Illumination and contrast can beimproved by regulation of light source and adjustment of condenser. In addition, staining the specimen with dyes may also increase contrast.

8.2.2 Microscopic Measurement

The size of a cell is one of important characters of a cell. As the cells are very small, they can only beobserved and measured under microscope. The actual size of an object should be expressed in micrometer (μm) not magnification. There are two devices to be used to measure the size of a cell, one is ocular micrometer which is placed into one ocular lens, and the other is stage micrometer. Ocular micrometer is a glass disk with a scale inthe center of the surface (Figure 8-2, left). A typical scale is composed of 50 or 100 divisions. To measure the length of an object under a microscope, count the number of the ocular divisions that span the object, and multiply by the conversion factor (the length of ocular division under a particular objective). As the magnification is different under different objectives, the conversion factor is also different. Therefore, it is necessary to calibrate the ocular scale by a stage micrometer at specific objective to get a conversion factor (a relative length of each ocular division) (Figure 8-3).

A stage micrometer is a microscope slide with a scale in the center of its surface (Figure 8-2, right). A typical micrometer scale is 1 mm long and divided into 100 divisions. Each division is 0.01 mm (10 μm).

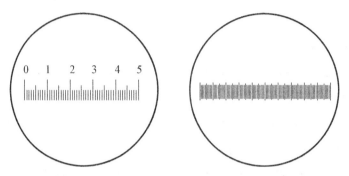

Figure 8-2 Ocular micrometer (left) and stage micrometer (right)

Figure 8-3 Calibration of ocular micrometer

8.2.3 Cell Cycle

Cell cycle includes cell growth and cell division, during which the hereditary materials are duplicated and divided into two daughter cells. The time of cell cycle is mainly spent in interphase (the process of cell growth) and the cell division is only a brief portion. Cell division occurs in both prokaryotes and eukaryotes, but differently. In eukaryotes, cell division usually involves two processes, mitosis (a process of nuclear division) andcytokinesis (a process of cytoplasmic division). The continuous process of mitosis is generally divided into five stages: prophase, prometaphase, metaphase, anaphase and telophase, based on the change of chromosomes.

(1) **Interphase:** The stage that cell is not dividing but synthesize some important materials, including proteins and DNA. The nucleus can be observedclearly with one or more dark nucleoli.

(2) **Prophase:** prophase is the first sign of mitosis, in which a thickening of the chromatin threads occurs. As prophase continues, the chromatids continue to condense into

chromosomes. In late prophase (also called prometaphase), the nuclear envelope and nucleoli disappear.

(3) Metaphase: At metaphase, the chromosomes align at the center of the spindle, which is also called the equatorial plate.

(4) Anaphase: The centromeresdivide and the separated sister chromatids move toward opposite spindle poles.

(5) Telophase: The two sets of daughter chromosomes at opposite spindle poles gradually uncoil and the nucleoli and nuclear envelope reappear. Cytokinesis may occur.

(6) Cytokinesis: The cytoplasm is divided at equator plate to form two new daughter cells, each with one nucleus.

In higher plants the process of forming new cells is restricted to special growth regions called meristems. These regions usually occur at the tips of stems or roots. In animals, cell division occurs almost anywhere as new cells are formed or as new cells replace old ones. Some tissues in both plants and animals, however, rarely divide once the organism is mature.

To study the stages of mitosis, you need to look for tissues where there are many cells in the process of mitosis. This restricts your search to the tips of growing plants such as the onion root tip (Figure 8-4) or, in the case of animals, to developing embryos such as the zygotes of horse roundworm (Figure 8-5).

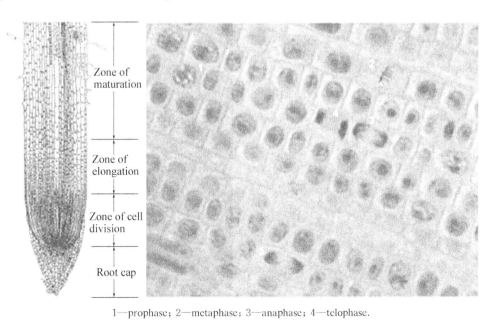

1—prophase; 2—metaphase; 3—anaphase; 4—telophase.

Figure 8-4 Longitudinal section of an onion root tip and mitosis

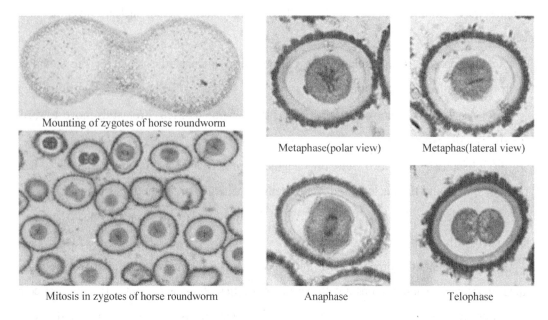

Mounting of zygotes of horse roundworm

Metaphase(polar view)

Metaphas(lateral view)

Mitosis in zygotes of horse roundworm

Anaphase

Telophase

Figure 8-5　Mitosis in zygotes of horse roundworm

8.3　Procedures

8.3.1　Standard Protocols for Operating Microscope

(1) General Notes

Please take good care in handling the microscope.

① Always carry the microscope with both hands.

② Always start and end with the lowest power objective.

③ Always clean the lenses with lens paper.

④ Always lower the stage completely, wrap electric cords and cover microscopes before returning them to the cabinet. Microscopes should be stored with the lowest objective in place.

(2) Focusing Specimen

① Clean the oculars and objectives with a piece of lens paper.

② Lower stage completely and make the lowest power objective in place.

③ Place the specimen on the stage.

④ Adjust the light intensity with light regulator switch and the condenser.

⑤ Use the focus wheels (first use the larger knob called coarse knob, and then use the small knob called fine knob) to focus, image may be small at this magnification.

⑥ Switch to low power objective and use the fine knob to refocus, and then to high power.

(3) Drawing Specimen

① Use pencil—you can erase and shade areas

② All drawings should be titled and labeled. The titleshould bewrittenat the bottom. The structure of a cell should be labeled clearly onthe rightside using straight lines which should never cross. The image scale is also required.

③ Shade areas should be densely or sparsely dotted with pencil carefully, not comma or dot with tail.

8.3.2 Observe word "a" slide

Take one "a" word slide, first, observe its size and location on the slidewithnaked eyes. After that, put the "a" slide under the lowestpower objective, pay attention to the difference between the direction of movement of slide and that of image.

8.3.3 Observe Different Morphological Characteristics of Cells from Different Tissues

8.3.3.1 Frog Renal Tubule Epithelial Cells

Observe the slide under the lowestpower objective. The renal tubules are comprised of single layered epithelial cells, which generate narrow tubes. The nucleus of the renal tubule epithelial cells is round and appear in dark blue. There are many dark granules in cell cytoplasm. Specially pay attention to the cell size and microscope magnification rates. Do not mistake the cell nucleus forthe cell.

8.3.3.2 Neuron Cells

Under the lowestpower objective, the tissue with the "butterfly" shape in the cross section of the spinal cord is the grey substance. Neuron cells are "star"-like cells in the grey substance area. The neuron cells are silver stained into brown color. Under high amplification lens, the neuron cells have many different sized spikes, with round shaped nucleus and nucleoli. The nucleoli were stained brown, and the nucleus was not stained.

8.3.4 Make mouth mucous membrane epithelial cells specimen

(1) Place a small drop of Trypan Blue on the center of a clean slide.

(2) Gently scrape the inside of you cheek with the broad end of a sterilized toothpick.

(3) Stir the scraped epithelial cells in the drop of Trypan Blue on the slide.

(4) Cover the drop with a coverslip lowered onto the slide at an angle to avoid forming air bubbles.

(5) Dry liquid off at the edge of the coverslip with a paper towel.

(6) Observe the slide under a microscope, first at low power and then at high power.

(7) Draw the appearance of your epithelial cells.

8.3.5 Measure the size of epithelia cells

(1) Calibrate the ocular micrometer scale using a stage micrometer

① Place a stage micrometer on the stage of a microscope.

② Rotate the eyepiece tube to make the ocular scale lined up with the scale of the stage micrometer.

③ Note the number of ocular divisions and stage division that cover a same distance, respectively.

④ Calculate the length of each division (conversion factor) of ocular micrometer based on the formula below.

$$\text{Conversion factor} = A \times B / C$$

A = length of each division of stage micrometer;

B = number of stage micrometer divisions;

C = correspondence number of ocular micrometer divisions.

(2) Record the number of ocular divisions that cover a cell (width and length) at low ($10\times$) and high ($40\times$) power, respectively.

(3) Then, multiply by the conversion factor for the magnification used.

(4) Measure at least three cells.

8.3.6 Observe chromosome behavior characteristics during different phases of cell cycle

Examine prepared slides of both onion root tips and roundworm blastula. Locate the meristematic region of the onion, or locate the zygotes with the 10X objective, and then use the 40× objective toobserveindividual cells.

(1) Observe and draw cells in different phases of mitosis.

(2) Try to consolidate the distinctive features of each phase.

(3) Try to estimate the number of chromosomes of this organism.

(4) Find a cell undergoing cytokinesis.

8.4 Results and analysis

(1) Observation of human mouth mucous membrane epithelial cells

Draw the epithelial cells you observed and label the nucleus, cytoplasm and cell membrane.

(2) Micro-scale measurement of cell size

① Calibratethe ocular micrometer at low power (10×) and high power (40×)(Table 8-1);

Table 8-1 Conversion factor at different magnification

eyepiece	Number of ocular Micrometerdivisions	Numberof stage Micrometerdivisions	Conversion factor(μm)
10×			
40×			

② Calculate the size of epithelial cells and fill in the following form(Table 8-2);

Table 8-2 Measurementsof the size ofhuman mouth mucous membrane epithelial cells

	1	2	3	average
length				
width				

(3) Calculation of results

Length(μm) = average number of divisions × conversion factor

Width(μm) = average number of divisions × conversion factor

The size of a cell: width (μm) × length (μm).

(4) Draw cells in different stages of cell cycle including interphase, prophase, metaphase, anaphase, telophase and cytokinesis.

8.5 Reagents

(1) Cedar oil.

(2) Dimethylbenzene.

(3) Letter "a" mounting (ready).

(4) Blue-red wool mounting (ready).

(5) Cut sections of frog kidney (ready).

(6) Transverse section of frog spinal cord (ready).

(7) Smooth muscle cell mounting (ready).

(8) Mitosis mounting of Cells ($2n = 16$) of root tip of onion (ready).

(9) Mitosis mounting of fertilized eggs ($2n = 4$) of horse roundworm (ready).

8.6 Instruments

(1) Ordinary optical microscopes.

(2) Pencil.

(3) Lens paper.

(4) Absorbent paper.

8.7 Trouble shootings

Occasionally you may have trouble working your microscope. Here are some common problems and solutions.

(1) Image is too dark

Make sure your light is on and adjust the condenser.

(2) There's a spot in my viewing field, even when I move the slide the spot stays in the same place

Your lens is dirty. Use lens paper, and only lens paper to carefully clean the objective and ocular lens. The ocular lens can be removed to clean the inside.

(3) I can't see anything under high power

Remember the step, if you can't focus underthe lowest powerand then low power, you won't be able to focus anything under high power.

(4) Only half of my viewing field is lit, it looks like there's a half-moon in there

You probably don't have your objective fully clicked into place.

8.8　Thought Questions

(1) How to distinguish between low, high power lens and oil immersion lens?

(2) How to adjust the brightness of objective image?

(3) What should you do after using microscope?

(4) How to judge where the dirt image of vision field is located? On slide or on eyepiece?

(5) How to tell animal cell mitosis from plant cell mitosis?

(6) How to define the relationship between cell mitosis and cytokinesis?

(7) Which part of the root ischosen for observing mitosis? Why?

(8) Which type of cytokinesis takes place in plant and animal cells?

Experiment 9

Cell Culture and Cell Fusion

9.1 Objectives

(1) Learn basic techniques of cell culture.
(2) Be able to count cells with hemacytometer.
(3) Understand the principle of cell fusion induced by PEG.
(4) Grasp the cell fusion technique.

9.2 Principles

9.2.1 Cell Culture

Cell culture is a kind of technique by which the cells (eukaryotic or prokaryotic cells) grow under specific conditions in vitro. For animal cell culture, it consists of thawing cell, growing cells, harvesting cells from dishes, splitting and plating cells on fresh dishes and storing cells by freezing.

When the cell is not used for experimental analysis, it is usually stored in liquid nitrogen. Dimethyl sulfoxide (DMSO) is a small molecular and easy-soluble protector when cell suffers freezing. To protect the cells from breaking up when meeting sudden decrease of temperature, it is necessary to put cells in freezing container and immediately into first at $-20°C$ then at $-80°C$ at last into liquid nitrogen at $-196°C$. The 10% DMSO and 10% fetal bovine serum (FBS) in basal medium can serve as a good cell freezing medium. However, at room temperature, DMSO is harmful to cells. Therefore, when thawing cells, we must immediately put the cells into $37°C$ to let the cells quickly thawed and get rid of DMSO.

(Slow freezing, fast thawing)

Cell culture can be divided into two types, primary cell culture and continuous cell culture. For primary cell culture, the cells are directly taken from the tissues of organisms and then cultured in suitable culture medium. For continuous cell culture, the cells are established stable cell lines, which are isolated from cancerous tissues or tissues containing stem cells.

The cell culture mediumcan be classified into two categories: defined mediumwhose composition is known and undefined media whose composition is unknown. The most commonly used cell culture media contains a kind of basal media (defined media), fetal bovine serum (FBS) and some other components to support the cell growth. FBS is derivedfrom the fetuses of cows and its composition is not known clearly.

9.2.2 Cell counting with hemacytometer

Hemacytometer is a thick glass microscope slide with two separate countingchambers. Each chamber has nine large squares and thelarge square in the corners has 16 smaller squares (Figure 9-1). After covered by a special coverslip, the volume above each large square is equal to 10^{-4} mm^3 (0.1 mL). So, we can calculate the concentration of cells in a cell suspension through counting the number of cells in the chamber of hemacytometer.

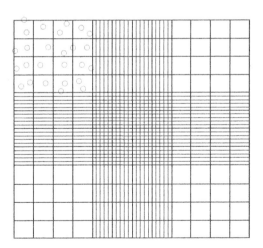

Figure 9-1 Hemocytometer under light microscope

9.2.3 Cell fusion

Cell fusion is also called cell hybridization, referring to the process of two or more cells fused into one heterozygote cell. Cell fusion is a multi-step process including cell membrane attachment and fusion, cytoplasm combination, cell nucleus, cell organelles and enzymes forming into one new complex system. As a useful technique, cell fusion is widely applied in cell biology. Monoclonal technique is a good example based on cell fusion, which contributes much to the life science research and medical application.

Polyethylene glycol (PEG) is a complex chemical jointed by many molecules of ethylene glycol, serving as a chemical pro-fusion reagent. PEG can combine with water through hydrogen bond and decrease the free water between cells, eventually destroy the phospholipid bilayer of cell membrane. The changed cell membrane structure makes cells easier to contact and fuseinto one. To improve the efficiency of PEG, it is necessary to use high concentrated PEG solution. On the other side, high concentration of PEG will do harm to the cells due to the over loss of water. Therefore, it is important toselect the proper molecule, concentration and working time of PEG to get the best result in fusion.

9.3 Procedures

9.3.1 Cell culture

(1) Cell thawing

① Preparation: Set up the water bath at 38℃~40℃ and prepare one centrifuge tube containing 9 mL medium with 10% FBS in the incubator.

② Thaw the cells: Take one tube of cells from the liquid nitrogen tan and immediately put it into the water bath. Shake softly to make it thaw quickly.

③ Transfer the thawed cells into the ready centrifuge tube in the flow cabinet. Centrifuge at 1 000 rpm/min for 5 min and discard the supernatant.

④ Re-suspend the cells with 5 mL fresh culture medium. Leave 0.5 mL for cell viability determination and the residuecells areculturedin cell cultureflask in the incubator at 37℃, with 5% CO_2.

(2) Cell splitting

For the adherent cells:

① Discard the supernatant medium and wash cells with PBS once.

② Add 0.25% trypsin-EDTA and incubate at 37℃ for 3~5 min.

③ Add 2~4 mL cell culture medium with 10% FBS to stop trypsinization.

④ Pipet up and down several times to break up the cell chunks.

⑤ Collect the cell suspension in a centrifuge tube and centrifuge at 1 000 rpm/min for 5 min.

⑥ Discard the supernatant, and re-suspend the cell pellets in desired volume of cell culture medium.

⑦ Replate the cells on cell culture dishes.

For the suspended cells: Directly collect the cell solution into the centrifuge tube and centrifuged at 1 000 rpm/min for 5 min.

(3) Store cells in liquid nitrogen

① Prepare the freezing working medium:

② Mix the serum-free 1640 basal medium, fetal bovine serum (FBS) and DMSO at a volume proportion of 8 : 1 : 1 to get a freezing working solution with 10% FBS and 10% DMSO.

③ Collect cells as showed in cell splitting.

④ Discard the supernatant and add freezing solution into the collected cells. Adjust the cell concentration to the final one of $(1\sim2)\times10^6$. Freeze the cells in aliquots with freezing tube and label carefully the tube including the cell line name, freezing date, cell concentration and operator name.

⑤ Freeze the cells: Leave the aliquots of the freezing tube of cells in the freezing container and put the container in $-80°C$ overnight. Then transfer the container into the liquid nitrogen tan for long storing.

9.3.2 Cell fusion

(1) Collect Hela cells by trypsin, centrifuge at 200 rpm/min and discard the supernatant and resuspend in 2~5 mL Hanks solution.

(2) Use the cell counter to calculate the blood cell concentration and Hela cells respectively. If the concentration is too high, dilute with Hanks solution to a final concentration of $(0.5\sim1)\times10^6/mL$.

(3) Mix well of the Hela cells and the chick blood red cells at ratio of 1 : 5(number : number), centrifuge at 200 rpm/min, discard the supernatant and put the tube in a water bath at 37°C (39°C is better) to pre-warm.

(4) Take 0.5 g PEG, heat on the alcohol burner to melt it, then cool down and add pre-

warmed 0.5 mL Hanks solution

(5) Add 0.5 mL of 50% PEG solution to the mixed cell (let the fusion reagent slowly flush down the tube wall), shake while adding, then leave it in the water bath at 37℃ for 5 min.

(6) Terminate the fusion process with 5 mL RPMI1640 medium without FBS.

(7) Centrifuge at 200 rpm/min and discard the supernatant.

(8) Resuspend the cells in RPMI1640 medium with 10% FBS and continue to incubate in the water bathat 37℃.

(9) Take some solution at different time points (5 min, 10 min and 15 min) for observation of cell fusion under the microscope.

Notes:

① DMSO is toxic to cells at room temperature (RT), therefore it is important to put the cells into the freezing temperature to avoid the toxicity to cells. Similarly, it is required to thaw the cells from the liquid nitrogenquickly.

② Factors that affect the cell fusion efficiency is multiple, among which, requirement for PEG is very strict. It is better to choose the PEG with molecular weight between 1 500~6 000 according to different cell types, which can be found in some references. The concentration of PEG is very important, better to set at 50%. The quality of PEG plays a pivotal role in cell fusion. So, it is better to use the imported type, but not the outdated one.

③ Cell fusion is very sensitive to temperature. It is not good for cell fusion at too high or too low temperature. The most favorite temperature varies between 37℃~39℃.

④ pH is also a vital factor which affects the cell fusion efficiency much. All the solution used in this experiment should be adjusted at the pH of 7.0~7.2.

⑤ To better observe the cell fusion process, cell staining should be performed, such as HE, Giemsa staining and Janus Green staining.

9.4 Results and analysis

(1) Observe the cell fusion process under the microscope and calculate the cell fusion ratio;

Cell fusion ratio = 100% × total cell nuclei involving in cell fusion in one view field/total cell nuclei in the same view field

(2) Draw the processof cell fusion.

9.5 Reagents

(1) Materials: HeLa cell line cultured in vitro; fresh chick blood;
(2) Chemical Reagents:
  ① PBS solution:

NaCl	7.9 g;
KCl	0.20 g;
Na_2HPO_4	1.44 g;
K_2HPO_4	1.8 g;
dd H_2O	800 mL

up to   1 000 mL

Adjust pH to 7.2~7.4

  ② Hanks solution:

Solution A:

$Na_2HPO \cdot 2H_2O$	0.06 g
KH_2PO_4	0.06 g
$MgSO_4 \cdot 7H_2O$	0.20 g
Glucose	1 g
NaCl	8 g
ddH_2O	750 mL

Solution B:

$CaCl_2$	0.14 g
ddH_2O	100 mL

Add solution B to solution A, and adjust the pH to 7.4 with $NaHCO_3$.

9.6 Instruments

Flow cabinet, incubator, centrifuge, light microscope, phase contrast microscope, liquid nitrogen tan, alcohol lamp, marker pens, cell flasks, measuring pipette, freezing tubes, weighing balance, water bath, hemacytometer, dropper, 10 mL centrifuge tubes, coverslips, slides.

9.7 Thought Questions

(1) Why is DMSO used for freezing cells?

(2) What vital steps should be paid special attention to when doing cell fusion experiment?

(3) How cell fusion is applied in monoclonal antibody?

Experiment 10

Primary Culture and Directional Differentiation of Rat Bone Marrow Mesenchymal Stem Cells

10.1 Objectives

(1) Master the basic methods and procedures for primary culture and passage of rat bone marrow mesenchymal stem cells.

(2) Master the aseptic technique in cell culture procedure.

(3) Learn the method for directional differentiation of rat bone marrow mesenchymal stem cells into adipocytes.

(4) Be familiar with the commonly used method for identifying the adipocytes derived from mesenchymal stem cells.

10.2 Principle

Stem cells are a group of cells which have self-renewal capacity and ability to differentiate into a variety of cell types under certain conditions. They can be found in embryos and adult tissues. In a developing embryo, stem cells can differentiate into all the specialized cells: ectoderm, endoderm and mesoderm. In adult organisms, stem cells and progenitor cells act to bea repair system for the body and toreplenish adult tissues. Stem cells can be cultured, passaged, cryopreserved and induced to differentiate into various types of cells in vitro, thus providing an ideal model to do research on biological science, such as human and animal embryogenesis, tissue cell differentiation and gene expression.

Furthermore, stem cells are potential candidates for cell replacement therapies to clinical tissue deficiency diseases and hereditary diseases. Stem cells are usually divided into embryonic stem cell and tissue stem cell according to their development stages, or into five categories based on their differentiation potency.

(1) Totipotent stem cells: which can differentiate into embryonic and extraembryonic cell types. Cells produced by the first few divisions of the fertilized egg are always totipotent.

(2) Pluripotent stem cells: which are the descendants of totipotent cells, can differentiate into nearly all types of cells derived from any of the three germ layers.

(3) Multipotent stem cells: which can differentiate into a number of cell types, but only those of a closely related lineage.

(4) Oligopotent stem cells: which can differentiate into only a few cell types, such as lymphoid or myeloid stem cells.

(5) Unipotent stem cells: which can produce only one cell type, but have the property of self-renewal, which distinguishes them from non-stem cells. Most tissue stem cells are lineage-restricted (multipotent) and are generally referred to by their tissue origin such as bone marrow mesenchymal stem cell (BMSCs), adipose-derived stem cell, etc.. The use of tissue stem cells in research and therapy is not as controversial as the use of embryonic stem cells, because the production of tissue stem cells does not require the destruction of an embryo. Additionally, tissue stem cells are obtained from the intended recipient; therefore, the risk of rejection is essentially non-existent.

Bone marrow is a rich source of tissue stem cells, besides hemopoietic stem cells, a great deal of research has been done on BMSCs, which are derived from mesoderm and have higher proliferation rate. BMSCs possess great plasticity and are able to differentiate into a variety of cell types, including osteocytes, chondrocytes, adipocytes, muscle cells and fibroblast cells. Recent studies showed that the etiology of osteoporosis and steroid-induced osteonecrosis is the increased formation of adipocytes in bone marrow. Fat cells are derived from pre-adipocytes coming from BMSCs. In this research, we try to establish the method to differentiate BMSCs into adipocytes in vitro, and thus providing a cell model to study the pathogenesis and preventive medicine of osteoporosis, steroid-induced osteonecrosis and obesity.

10.3 Procedures

10.3.1 Primary culture of rat BMSCs

(1) Reagent preparation

① BMSC culture medium: DMEM/low glucose + 10% FBS + penicillin/streptomycin.

② PBS: 8.0 g NaCl, 0.2 g KCl, 1.56 g $Na_2HPO_4 \cdot 2H_2O$, 0.2 g KH_2PO_4 are dissolved in 1 L H_2O, autoclaved and stored at 4℃.

(2) The culture medium is pre-warmed at 37℃.

(3) UV irradiationis turned on to sterilize biological safety cabinet for 15 min; the operation table is cleaned with 70% ethanol.

(4) The required materials and reagents are put in the cabinet (spray each item with 70% ethanol and swab with autoclaved paper).

(5) Pre-cooled PBS plus penicillin/streptomycin is added into three 10 cm culture dishes, and DMEM/low glucose into another dish.

(6) The rat is sacrificed and immersed in 70% ethanol for 1 min, then transferred into the dissecting pan in the cabinet, the skin is opened to exposure femur and tibia, which are collected and transferred to pre-cooled PBS in dish, vascular and connective tissue are removed and the bone is transferred to another dish with PBS for washing.

(7) The end of bone is cut off, the bone marrow is aspirated with DMEM/low glucose using 30 g syringe and separated into single cells by pipetting several times. Cells are centrifuged at 1 500 rpm, 10 min.

(8) Cell deposition is resuspended and seeded into 25 cm^2 culture flask at a cell density of 5×10^7 per flask or into 6-well culture plate with pre-coated cover slide.

(9) 24 hours later, the medium is changed and cells are washed with PBS twice to remove non-attached cells.

(10) Medium is changed every third day.

(11) The BMSCs will grow to 80% confluence within 7~10 days, the cells are dissociated with 0.25% trypsin-EDTA and passaged at a ratio of 1:3 or cryopreserved in DMEM/low glucose + 30% FBS + 5% DMSO at a cell density of 1×10^6 cells/mL.

10.3.2 Directional differentiation of BMSCs into adipocytes

(1) Reagent preparation

① 10 mg/mL poly-lysine stock solution: 10mg poly-lysine is dissolved in 1 mL H_2O and filtered with 0.2 μm filter. Working solution: 400~500 μg/mL.

② Adipogenic induction medium: DMEM/low glucose supplemented with 10% FBS, 10^{-6} M dexamethasone, 10 μg/mL insulin, 0.5 mm 3-isobutyl-1- methyl-xanthine (IBMX), 0.2 mm indomethacin, 100 U/mL penicillin and 100 μg/mL streptomycin.

(2) Coating cover slide: put the cover slide into 6-well culture plate, add poly-lysine working solution into plate, 3~4 hours later, remove poly-lysine and wash the cover slide with H_2O for 3 4 times, dry culture plate in incubator.

(3) BMSCs at passage 2 are cultured on cover slide in 6-well culture plates at a cell density of 1×10^5 cells/well. When cells grow to 80% confluence, medium is changed with adipogenic induction medium. Medium is change every third day. The cells are stained with oil-red-O at days 7 and 14 post-induction, observe the cells under microscope and take pictures.

10.3.3 Oil-red-O staining and adipocyte counting

(1) Oil-red O stock solution preparation: 0.4 g oil-red O is dissolved in 10 mL isopropanol and stored at room temperature. Working solution: stock solution is mixed with H_2O at a ratio of 3∶2 and filtered with filter paper.

(2) Cells are washed with PBS for 1 min and fixed with 10% formaldehyde for 10 min, and then stained with oil-red O working solution for 10 min. Cells are washed with 60% isopropanol to remove redundant staining solution and further washed with PBS three times, 1 minute each time.

(3) Observe the cells under microscope, count oil-red-O stained cells in 10 random fields, and calculate the differentiation ratio.

10.4 Results and Analysis

Take pictures and describe the characteristics of BMSCs and differentiated cells.

10.5 Reagents

(1) 2-month-old SD rats

(2) PBS, fetal bovine serum (Biowest), 0.25% trypsin-EDTA, 70% ethanol, poly-lysine, 10% formaldehyde, DMEM/low glucose, dexamethasone, insulin, indomethacin, 3-isobutyl-1-methyl-xanthine

(3) Insulin stock solution (1×10^4 mg/L): 25mg insulin (Sigma, I5500) is dissolved in 2.5 mL hydrochloric acid (pH=2), filtered and stored at $-20°C$ for one month. Working concentration is 10 mg/L (Dilute 1 000 times).

(4) Dexamethasone stock solution (1×10^{-3} m): 3.9246 mg dexamethasone (Sigma, D1756, molecular weight 392.46) is dissolved in 10 mL ethanol, filtered and stored at $-20°C$. Working concentration is 10^{-7}M (Dilute 10 000 times).

(5) Indomethacin stock solution (0.2 m): 751.58 mg indomethacin (Sigma, I7378, molecular weight 357.7) is dissolved in 10 mL DMSO, filtered and stored at $-20°C$. Working concentration is 0.2 mm (Dilute 1 000 times).

(6) IBMX stock solution (0.5 m): 250 mg IBMX (Sigma, I5879, molecular weight 222.24) is dissolved in 2.249 8 mL ethanol, filtered and stored at $-20°C$. Working concentration is 0.5 mm (Dilute 1,000 times).

(7) 10% formaldehyde: 120 mL formaldehyde, 4 g $NaH_2PO_4 \cdot H_2O$, 13 g Na_2HPO_4, 880 mL H_2O, filtered and stored at 4°C.

10.6 Instruments

Dissecting instrument, gauze, 15 mL and 50 mL tube, cell culture dish, cell culture flask, 10 200 μL and 1 000 μL tip, pipette, slide, cover slide, cell culture plate, straw, shaker, oven

10.7 Thought Questions

(1) How to identify stem cells in cultured cells?

(2) To describe the different aspects of embryonic stem cells and tissue stem cells.

Reference

[1] Hou LL, Cao H, Zhang Y, et al. Progress on the Study of Tissue Stem Cells[J]. Journal of Experimental Hematology. 2002,10(2):159-162.

[2] Sekiya I, Larson BL, Vuoristo JT, et al. Adipogenic differentiation of human adult stem cells from bone marrow stroma (MSCs)[J]. Bone Miner Res, 2004,19(2):256-64.

[3] Li XH, Zhang JC, Sui SF, et al. Effect of daidzin, genistin, and glycitin on osteogenic and adipogenic differentiation of bone marrow stromal cells and adipocytictransdifferentiation of osteoblasts[J]. Acta Pharmacol Sin, 2005, 26(9): 1081-1086.

[4] Li Jiao, Shi Xiujuan, Wang Juan, et al. Comprehensive Experiment-Clinical Biochemistry: Determination of Blood Glucose and Triglycerides in Normal and Diabetic Rats[J]. Biochemistry and Molecular Biology Education, 2015, 43(1):47-51.

[5] Vehik, K., Ajami, N. J., Hadley, D., et al. The changing landscape of type 1 diabetes: recent developments and future frontiers[J]. Curr Diab Rep, 2013, 13, 642-650.

[6] Deeds, M. C., Anderson, J. M., Armstrong, A. S., et al. Single dose streptozotocin induced diabetes: considerations for study design in islet transplantation models[J]. Lab. Anim, 2011, 45(1), 131-140.

[7] Shek, P. N., Howe, S. A. A novel method for the rapid bleeding of rats from the tail vein[J]. J Immunol Methods, 1982, 53(2), 255-260.

[8] Lv LX, Li J, Shao ZH, et al. Practice of "integrated life science" curriculum[J]. China Higher Medical Education, 2012(4):79-80.